Working Nature

Daniela Russ is a historical sociologist based at the Global and European Studies Institute of the University of Leipzig, Germany. Her research explores the history of the energy economy in a global perspective, the conflicts around the decarbonization of the electric grid, and the theory and practice of Soviet energy planning. Her work has been published by outlets such as *Contemporary European History, Historical Materialism,* and Stanford University Press.

Working Nature

A History of the Energy Economy

Daniela Russ

London • New York

First published by Verso 2026

The manufacturer's authorized representative in the EU for product safety (GPSR) is LOGOS EUROPE, 9 rue Nicolas Poussin, 17000, La Rochelle, France
contact@logoseurope.eu

1 3 5 7 9 10 8 6 4 2

Verso
UK: 6 Meard Street, London W1F 0EG
US: 207 East 32nd Street, New York, NY 10016
versobooks.com

Verso is the imprint of New Left Books

ISBN-13: 978-1-80429-897-8
ISBN-13: 978-1-80429-898-5 (UK EBK)
ISBN-13: 978-1-80429-899-2 (US EBK)

British Library Cataloguing in Publication Data
A catalogue record for this book is available from the British Library

Library of Congress Cataloging-in-Publication Data
A catalog record for this book is available from the Library of Congress

Typeset in Minion by Hewer Text UK Ltd, Edinburgh
Printed and bound by CPI Group (UK) Ltd, Croydon CR0 4YY

Contents

Introduction

A wish for unification has always stood at the centre of productivism: that nature's purpose and society's might converge in an economy of energy. In 1928, the Soviet geophysicist Boris Veinberg published a short essay, 'The Conquest of Power', in the popular Soviet journal *The Universe and Humanity*. The article advanced a broad history of energetic progress, dating from its neolithic origins to a future where its development would enable humanity's practical omnipotence. Veinberg, who spent most of his life in the Siberian city of Tomsk, wrote on the margins of industrial modernity, but he was well travelled and imaginative. A natural scientist, he offered readers a rich anthropological account of the development of material culture: Through the ages, human beings have learned to create, conserve, move, and transmit force, light, heat, sound, and matter. This gradual progress leads to a 'unified world economy' covering the entire globe, in which 'energy' will have become as natural an object of consumption 'as food and air are now'.[1] Veinberg's text echoes the idea, expressed in so many languages at the time, that progress is materially rooted and that it depends on the control and appropriation of natural forces.

1 Boris P. Veinberg, 'Zavoevanie moshchnosti', *Vselennaia i chelovechestvo* 12 (1928): 702–5.

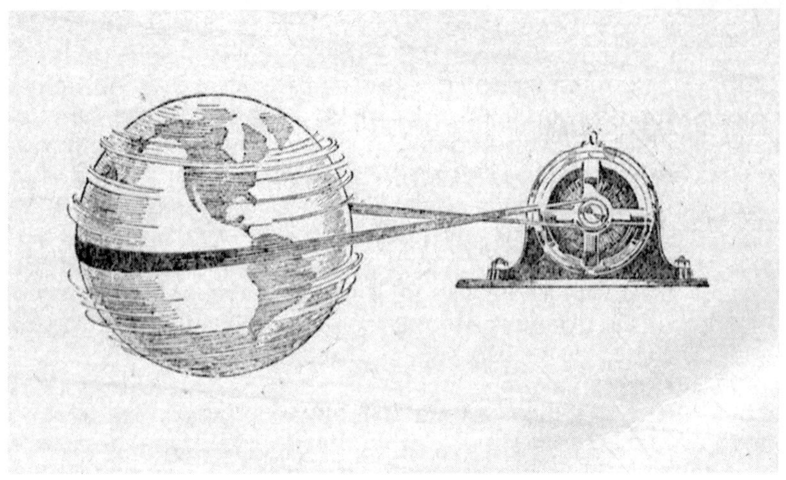

Figure 0.1. Boris P. Veinberg, 'Zavoevanie moshchnosti' [The conquest of power], *Vselennaia i chelovechestvo* 12 (1928): 706

Such discussion was conventional. But a small illustration accompanying the essay, depicting a planet connected to a generator – and appearing without title or credit name – can be read against such a convention. This image stands out from those scattered throughout the text. The first page includes a socialist realist masculine figure steering the wheel of industry; other images are evocative of modern machinery and life. Yet the Planet-Generator is neither realist nor heroic, though it is perhaps unintentionally ironic. With the authority of a scientist, Veinberg writes that technical progress is leading towards an all-powerful humanity. That human control of nature was imminent must have sounded fantastic to a Soviet population that had just experienced another harvest failure. The image subtly undermines Veinberg's argument: It destabilizes the idea of harnessing nature for human purposes. Just as the viewer begins to trace the belts, the transmitters of energy, humanity's control of natural power begins to dissolve. The illustration oscillates between two unsettled yet mutually constitutive meanings. Does the earth drive the generator, or does the generator spin the earth?

From the vantage of Earth driving the machinery, the image articulates humanity's capacity for harnessing energy on a planetary scale. It symbolizes the end point of this conquest, the historical moment when not only biomass, fossil fuels, and solar radiation but also the mechanics of the universe itself will be put to work in human machines to serve human ends. Those belts transmitting movement terminate at the generator, the productive site where human thought apparently materializes and manifests itself as world-making force. Yet this machinery sits in a void, and no instrument of labour is attached to it – it produces nothing. Movement could very well continue for as long as the earth spins, dissipating heat into the universe, and altering nothing. After all, no history would be made.

The image, however, might just as well be read as a representation of Earth itself being worked. This reading would see in the image the symbol of a self-generating humanity consciously determining its means of reproduction, inclusive of its interaction with the very processes of the planet and the wider universe. If humanity could control the planet's angle and rotational speed, it could alter or halt the earth's seasons and its day–night cycle. It would take on a sun-like role in organizing the planetary system. Yet the reading is adequate only as a first interpretation; it evaporates when one realizes that the generator depicted has no input whatsoever. It was all just an idealist dream to make something out of nothing. Whichever way we read it, the illustration shows a collapsing materialism, an idealist end point appearing within an energo-materialist understanding of human history. Neither the natural conditions nor the historical purpose of this conquest of nature's power can be represented.

The Planet-Generator image uniquely illustrates the historical meaning of the concept of energy and offers a concentrated example of its contradictions. The energy *concept* is not synonymous with the word 'energy' but pertains to the idea of nature's capacity to perform work as it is manifested in a variety of words, units of measurement, and objects. To approach nature with the concept of energy in mind means to treat those objects which are phenomenologically radically distinct – rivers, human beings, wind, coal – as they relate to one another and are interchangeable in terms of their performance of work in production. A tension arises insofar as the concept emphasizes the reliance of human history on natural forces beyond itself as a species; while, at the same time, it suggests that humanity has grasped the natural foundation of its own development. This latter point is only one step removed from a claim that there is a discoverable single principle governing both natural and social change on which natural and social history converge. It follows that emancipation from nature is cast as mastering this principle of change – the very 'changeability' of things. At the height of enthusiasm for energetic thinking, dating to the turn of the twentieth century, energy was, simultaneously, the focus of the most materialist and idealist projections: It was the basis of all matter, and it was itself a unifying concept.

Past visions of domination, such as Veinberg's 'Conquest of Power', are all too easily reduced to objects of critique or even ridicule. But it is understandable that responses to the harshness of life might take on a utopianism of control, redirecting the instinct to dominate at what has dominated. This fantasy, which perceives nature as something to be conquered, contains an emancipatory quality in its wager that things could be very different, that mastery of nature could reduce human suffering. Might the idea of reconciling nature's malleability with human emancipation be worth rescuing – especially during a period of so-called energy transition? Alas, for the duration of Boris Veinberg's life, nature's powers were not primarily applied to the end of making human life easier. Little over a decade after arguing that the conquest of nature's power was imminent, he died of hunger – lacking the 'most natural object of consumption' – in a Leningrad besieged by a fascist enemy that directed weaponized energies against him. Might the illustration, in fact, capture the failure to reconcile nature and history, the divergence between the potential and reality of harnessing natural forces for the good life – a divergence that is even more acute today than in Veinberg's time?

Today, the ease with which we identify the strangeness, hubris, and tragedy of the early-twentieth-century energetic ambition disguises its continuation in other forms. Veinberg may have underestimated capitalism's capacity for globalizing energy production and distribution – to him, this was a task for a coordinated, rationalized socialist system of production. But he was correct in predicting that energy would become the most natural object of consumption. Today there is a world-encompassing energy economy, one that is technically and legally standardized, producing homogeneous commodities and similar forms of ownership. Primary energy sources such as oil and gas are among global capitalism's essential commodities, as are 'energy converters' such as cars and machinery. The global oil trade remains the world's largest commodity market and a key indicator of the state of the global economy, just as electricity consumption penetrates modern life so deeply that it can be used as a real-time gauge of the dynamism of the economy itself. Some argue that the global energy industry relies on the largest and most comprehensive infrastructure ever built. Its physical plant extends down ten kilometres under the earth's surface and is found in the open ocean, hundreds of kilometres off land. It generates energy across the world and outer space. Two centuries of technological history are embedded in its structures, and tens of trillions of dollars are invested in it.[2]

In time and space, then, the world is now more comprehensively assessed in energetic terms than ever. With ice sheets melting, geological surveys of the Arctic indicate that the region holds much of the world's global oil and gas reserves. Likewise, the seabed has now been mapped in great detail, in the hope of extracting its mineral-rich soil for a promised energy transition, or to the end of expanding offshore oil and gas drilling. Though still in the realm of speculation, some astrobiologists now predict a prospect of heavy oil (similar to the Canadian tar sands) on Mars and call for an exploration of the solar system as a 'universal unconventional petroleum system'.[3] With the exploitation of atmospheric pressure differences and electromagnetic waves, the energetic measurement of the globe has become total: Instead of pockets of energetic resources, we now

2 Karen C. Seto et al., 'Carbon Lock-In: Types, Causes, and Policy Implications', *Annual Review of Environment and Resources* 41, no. 1 (1 November 2016): 426; Vaclav Smil, *Energy Transitions: History, Requirements, Prospects* (Praeger, 2010), 13.

3 Prasanta K. Mukhopadhyay, David J. Mossman, and James M. Ehrman, 'The Case for Vestiges of Early Solar System Biota in Carbonaceous Chondrites: Petroleum Geochemical Snapshots and Possible Future Petroleum Prospects on Mars Expedition', in *Instruments, Methods, and Missions for Astrobiology X*, vol. 6694 (SPIE, 2007), 107–22.

have a continuous, increasingly fine-grained map of shades of energetic potential. Curves of oil field exploitation have always extended resource exploitation into the future; but wind and solar power forecasts could assign such a future potential to every discrete cell of global surface models per unit of time at ever-increasing spatial and temporal resolution. Energy measurements are integrated into daily life to an unprecedented degree, promising control over the minutiae of the human body and habitation: Digital technology tracks one's steps, modulates household appliance settings, and accounts for the calorific value of every morsel of food.

As society has subjected itself to more extensive and detailed measurements, history's narration has also been transformed. The energy-historical speculations of the nineteenth century have given way to the scientific study of past energy use and future ecological trajectories. By assessing energy regimes historically, ecologists and environmental historians have reinterpreted the use of fire, the agricultural revolution, and industrialization in terms of the flow, transformation, and storage of planetary energy.[4] For most of human history, aggregate energy use remained quite stable. 'Empires rose and fell . . . but the energy calculations remained much the same.'[5] The substitution of measurement (energy calculations) for the object measured (energy use) is a beautiful mistake, as it points to one central problem encountered when writing a history of energy. The calculations could not have 'remained the same', because they simply could not have been made at the time in question. No energy history is without a concept of energy. Energy calculations have required a painstaking reconstruction undertaken by later historians, for whom deep energy history had become a way of understanding the peculiarities of the energetic predicament of the present. This present is generally said to have begun with the Industrial Revolution.

4 Vaclav Smil, *Energy in World History* (Westview Press, 1994); Vaclav Smil, *Energy and Civilization: A History* (MIT Press, 2017); Alfred W. Crosby, *Children of the Sun: A History of Humanity's Unappeasable Appetite for Energy*, 1st ed. (W. W. Norton, 2006).

5 Edmund Burke, 'The Big Story: Human History, Energy Regimes, and the Environment', in Edmund Burke and Kenneth Pomeranz, eds, *The Environment and World History* (University of California Press, 2009), 38.

Reinterpreted as a transition to a fossil fuel energy regime, industrialization represented an unprecedented rupture in human relations with nature, which historians have documented meticulously.[6] More energy has been consumed in the period since 1945 alone than in the entire Holocene epoch.[7] Such an energetic history of human energy regimes can also be projected into the future: While there is widespread pessimism about humanity's capacity for controlling energy after phasing out fossil fuels, some Earth-system scientists maintain that humanity, by packing Earth's surface with photovoltaic panels, may in fact be able to capture *more* free energy on the planet.[8]

The 'humanity' of such historiographies, the agent of progress, never existed. The energy economy's expansion and technical homogeneity are permeated by stark social differences. Some people are primarily consumers of energy, and it is only when they are deprived of it that they are reminded of what it means to command energy at orders of magnitude greater than their own bodily strength. On the battlefield or in the factory, others experience daily the consequences of being subjected to external powers. Some workers must adapt their hands, eyes, and minds to the accelerated, digitized production lines, while others, cheaper to exploit and lacking the latest machinery, scrape coal with their bare hands. Then there are those who sell their intellectual labour, their skill of organization or modelling, to think through the energetics of production and to optimize it. Some live far from the mines where the raw material necessary for energy infrastructure is found and processed, while others struggle to survive among the tailings of factories – without ever benefiting from nature's own work. Lest we forget those who can summon nature's power for any spontaneous desire, be it a trip into space or a cruise around the

6 Astrid Kander, Paolo Malanima, and Paul Warde, *Power to the People: Energy in Europe over the Last Five Centuries* (Princeton University Press, 2014); Simon Pirani, *Burning Up: A Global History of Fossil Fuel Consumption* (Pluto Press, 2018).

7 Jaia Syvitski et al., 'Extraordinary Human Energy Consumption and Resultant Geological Impacts Beginning Around 1950 CE Initiated the Proposed Anthropocene Epoch', *Communications Earth and Environment* 1, no. 1 (16 October 2020): 1–13.

8 Axel Kleidon, *Thermodynamic Foundations of the Earth System* (Cambridge University Press, 2016), 347; Axel Kleidon, 'How the Technosphere Can Make the Earth More Active', *Technosphere Magazine* (blog), 29 May 2019.

world's oceans. Because concentration of control and ownership is traditionally high in all sectors of the energy economy, the decisions of only a small number of people will affect the conditions under which everybody else works and lives, sweats, and breathes.

There is good reason for the diminishing prominence of stories in which human history and control over energy move in step. In a period of climate change and general environmental degradation, the question is no longer who shares in progress, but whether one can speak of progress at all. Out of a rational, highly technical infrastructure that is the energy economy, a social relationship to nature emerged that has run out of control. It is no longer plausible to argue 'that human society and nature are linked by the primacy and identity of all productive activity' – as Anson Rabinbach famously put the energetic worldview. By realizing in practice an energetic relation to nature, and by controlling individual energies, it has become evident that the energy economy does not correspond to the mastery of the natural conditions of human life.

How to Write Energy History

The academic literature distinguishes between energy as natural law outside of history, and the cultural forms related to it, which change over time. From this perspective, natural scientists are responsible for studying energy in nature itself – as in the laws of thermodynamics – while historians and social scientists confine themselves to the study of the development of ideas, concepts, and related practices in energy's orbit. The field of environmental history, however, is distinct in its attempts to assess both: It starts from the ecological premise that nature is governed by energetic laws, and then asks how energy conditions, and is conditioned by, social change. Environmental history understands energy history as an 'underlying structural history', as Rolf Peter Sieferle once put it. Energy has influenced historical change unconsciously.[9] History, from the environmental-historical viewpoint,

9 Rolf Peter Sieferle, *Rückblick auf die Natur: Eine Geschichte des Menschen und seiner Umwelt* (Luchterhand, 1997), 32.

appears mainly as a succession of energy regimes. This 'ecologist' line of thinking holds that the energy system or regime of a society is embedded in nature; it partakes in nature's flows of energy and itself develops from it. Energy is society's material basis.

The present book takes yet another route into energy history. Written in the tradition of historical materialism, and influenced by the Frankfurt School specifically, it starts from an experience of nature as immanent: Nature is the concept for the practical experience *of that which is neither social nor cultural.* Mental, social, and cultural forms operate according to concepts, norms, and ideas. These forms always contain an experience of nature, of what they are not. This fact expresses itself in our relation to our own nature, to the mortality of our bodies, and to the question of *Bildung* and socialization (how one becomes a social subject). This experience of nature in social practice precedes scientific knowledge.[10]

Human labour also involves this experience of nature. 'Nature . . . can only be regarded as unformed material from the point of view of the purposes of human activity.' Because the stuff of nature is already formed according to natural regularities, human beings must labour to appropriate them 'in a form adapted to [their] own needs'. Only because the things in nature exhibit regularities can they be used for human purposes and be made into tools. Still, nature remains ultimately indifferent to human formations; all products of human labour are subject to decay. There remains a non-identity between the subject and object of labour, a source of resistance: Labour means the human struggle to shape natural things into products for human use against nature's resistance and to keep the human-made world from sinking back into nature. Production thus retains an experience of what is not productive, what cannot be tamed, what resists being shaped into useful forms.[11] Perhaps more than ever, today's ecological

10 Christoph Menke, 'Die Lücke in der Natur: Die Lehre der Anthropologie', in *Am Tag der Krise: Kolumnen* (August Verlag, 2018), 32–4.

11 Karl Marx, *Capital: A Critique of Political Economy*, vol. 1 (Penguin Classics, 1992), 283; Alfred Schmidt, *The Concept of Nature in Marx* (Verso, 2014), 63, 74–5.

questions compel us to reconsider nature's negation of human labour.[12]

A materialist history of the energy economy would understand the concept of energy, the idea of nature's capacity to perform work, in its relation to production. This relation has two components: First, the concept of energy emerges historically in production, in working nature. Once the idea is formed, however, it transforms production, changes how natural things are approached, and informs what is searched for in nature. Second, the stuff of nature is then shaped into things that can perform regular work – processed coal, refined oil, the electrical current, and so on. In this conscious reshaping of things into energy products, the concept becomes objectified, embodied in resources, commodities, machinery, and measuring devices. The world becomes more like the idea (but never quite). As human labour both negates and actualizes nature's forms, so does the concept of energy that is applied in this process.

Much energy and environmental history, including some eco-Marxist scholarship, pits the *nature* of energy against the *social form* of the economy. This is not to say that ecologist historiographies continue the work of the nineteenth-century energeticists like Ernest Solvay, Wilhelm Ostwald, or the above-mentioned Boris Veinberg, for whom there was an energetic *purpose* in nature that society must realize. Anthony Wrigley and Rolf Peter Sieferle responded to an entirely different historical situation and intellectual field. At a time when economic growth began to flag in the 1970s, they revisited energy's origins in the ecological conditions of the industrial revolution.[13] They borrowed their concept of 'materiality' from the science of ecology in order to argue that ongoing economic growth or capital accumulation, which knows no measure or end, will run up against natural limits of land use and thermodynamics. In so

12 To speak of nature's resistance does not imply a subjectivity or agency of nature: Nature resists or 'takes revenge' (in Engels's words) only from the point of view of human activity.

13 E. A. Wrigley, *Energy and the English Industrial Revolution* (Cambridge University Press, 2010); Rolf Peter Sieferle, *The Subterranean Forest: Energy Systems and the Industrial Revolution*, trans. Michael P. Osmann (White Horse Press, 2010).

doing, they introduced the fact of natural constraints not often considered by mainstream economics.

As long as capitalists claim to be the sole creators of wealth, it will be necessary to emphasize the natural conditions of production. But if one takes the ecological explanation sketched by Wrigley and Sieferle seriously, one must also assume a pre-given energetic *need* in society that meets an energetic *potential* offered by nature – however mediated this appears in their explanations by allusion to other social factors. Societies certainly draw on nature for the satisfaction of their needs, but that those needs should take the *form of energy* is a specific development requiring historical explanation. Is there an imperative to grow *in terms of energy* – and where does it originate? Ultimately, the ecological concept of nature remains incapable of understanding the origin and development of our current energetic predicament and cannot grasp its own complicity in it. The world-encompassing infrastructure described above did not develop *naturally* from nature's energy flows; no natural law leads from a nature governed by energetic laws to today's energy economy. Ecology can only describe, but never explain, this spectacularly *unnatural* development of the energy economy. To understand this history, we need to ask what energy is and was, not in nature or to the human species ecologically conceived, but to the people who put nature to work.

Historical materialism counters the assumption of a natural tendency towards growth by revealing the compulsion to increase productivity as an outcome of competitive production for the market specific to capitalism.[14] Material or energetic growth is then not a natural process, but the indirect outcome of production for maximization of surplus value. In their different ways, Elmar Altvater, Matt Huber, and Andreas Malm argue that the shift to fossil fuels – perhaps the best-studied moment in energy history – must be understood within the capitalist form that the relations of production had already taken in Great Britain.[15] As Andreas

14 Ellen Meiksins Wood, *The Origin of Capitalism: A Longer View*, rev. ed. (Verso, 2017).

15 Andreas Malm, *Fossil Capital: The Rise of Steam Power and the Roots of Global Warming* (Verso, 2016); Matthew T. Huber, 'Energizing Historical Materialism: Fossil Fuels, Space and the Capitalist Mode of Production', *Geoforum* 40, no. 1 (1

Malm shows in *Fossil Capital*, the energetic hypothesis is difficult to sustain even for its initial explanandum, the British Industrial Revolution. The shift from water to steam power in the textile industry was driven less by concerns about the availability and cost of energy than by the cost and discipline of labour (an argument that awaits its wider contextualization in the development of other sectors, such as the iron industry). One way to conceptualize this relationship – that steam cheapens labour – would be to say that the driving force of machinery functioned as a factor of production that interacted with land, capital, and labour. This raises the question of when nature's work became a factor of production and could be treated independently from land and labour.[16]

While not primarily concerned with questions of origins, this book ventures a working hypothesis about it. I will argue that the origin of the concept of energy – understood as not the first use of the term 'energy', but the emergence of the material conditions of experiencing and treating nature as working – lies in the objectification of the motor mechanism in large-scale industry. From the perspective of an industrial capitalist, all productive work – whether performed by human labour or natural forces – becomes identical, equally purchasable, independent of its concrete specificity. In comparing the costs of steam and water power, for instance, millowners were already abstracting from any particular natural force and treated the driving force as a more general function that could be performed by different sources of power. This does not by itself constitute an *imperative* to grow energetically. Rather, with an independent motor mechanism that can be substituted, the calculation of nature's work and its cost becomes a moment of capitalist competition and class conflict, even when millowners *refuse* cheaper or more powerful forms of energy. In short, I suggest that the concept of energy should be seen in relation to a real subsumption of production, as a derivative of the abstraction of labour.

Energy is the concept that mediates the appropriation of processes and materials for the performance of work in production. The energy

---

January 2009): 105–15; Elmar Altvater, 'The Social and Natural Environment of Fossil Capitalism', *Socialist Register* 43, no. 43 (2007): 37–49.

16 Altvater, 'The Social and Natural Environment of Fossil Capitalism', 41.

economy is therefore the sphere in which the prime mover of machinery is developed and improved, where natural materials and processes are utilized and reshaped to perform work. Unlike the category of 'energy system', which captures structural and technical changes in energy use over centuries, the energy economy is a historically particular category: There is no pre-capitalist, pre-industrial energy economy. In no other energy system have people consciously used a quantified amount of nature's changeability and bound their own emancipation to it; never before have they bought, sold, and invested in it. This book thus offers an entwined history of the energy economy as the becoming of the concept and the thing.

'Ecologist' environmental historians maintain that nature is ordered energetically, and that society partakes in this order. They understand today's energy regime as having developed from nature – not as human realization of nature's purpose, but as an energetic potential in nature that meets an energetic need of society. In contrast, the present book argues that the energy economy develops from nature not by actualizing and affirming an energetic order in nature; its history is not a movement to an ever-closer realization of the energy concept in nature, not a *positive* dialectic. Rather than assuming that the regularities of change in nature are the foundation of the energy economy, this book endeavours to understand the energy economy as a realization of society *against* resistance from nature. It asks how engineers, scientists, and economists accomplished the production and circulation of a working nature against social and natural resistance and how, by doing so, they also undermined the promise of the energy economy to deliver emancipation from nature.

A Critical Materialism

This suggests that one must conceptualize energy as a term opposed to nature to retain materialism. To understand energy as a concept emerging within and informing social practice is, however, neither a claim that society *produces* nature, that it makes nature as it wishes, nor a claim that it creates nature's capacity to work. It is no abandonment of materialism: Society objectifies itself in production, 'without however "positing" the objectivity of nature as such'.[17] But what would a materialist perspective of the field of energy be, when one no longer accepts the ecological view that nature is made up of flows of energy and matter? What does it mean to hold nature's objectivity against the concept of energy?

Frankfurt School 'critical materialism' – as articulated by Theodor Adorno and Alfred Schmidt – has often been misunderstood, and so its use here requires a brief clarification.[18] Adorno acknowledged the priority of matter over mind, of nature over culture, in what he called the 'primacy' or 'preponderance' of the object. In this view, the primacy of the object renders dialectics materialistic: There would be no dialectics in matter if it were 'total, undifferentiated, and flatly singular'.[19] Subject and object stand in an asymmetrical but not dualistic relationship to one another. The terms are intermediated, but in different ways: The 'object is also mediated; but . . . it is not so thoroughly dependent upon subject as subject is dependent upon objectivity'.[20] That there is no knowledge of an object without a conscious subject does not constitute an ontological primacy of the subject. It is the very meaning of objectivity to exist independently of a subject. Subjects, in contrast, are never subjects alone: They are also objective. Experience and practice – indeed, all striving for realizing emancipation in the world – would be impossible if the subject were radically different and did not belong '*a priori* to the same sphere' as the object.[21] Still, that subjects cannot be severed from their objectivity does not mean that they can be reduced to it. The affinity of subject and

17 Schmidt, *The Concept of Nature in Marx*, 65.

18 Alfred Schmidt, *Geschichte des Materialismus*, 1st ed. (Salier Verlag, 2017), 25; Matthias Lutz-Bachmann and Gunzelin Schmid Noerr, *Kritischer Materialismus: Zur Diskussion eines Materialismus der Praxis* (Hanser, 1991); Deborah Cook, 'Adorno's Critical Materialism', *Philosophy and Social Criticism* 32, no. 6 (September 2006): 719–37.

19 Theodor W. Adorno, *Negative Dialectics* (Routledge, 1973), 205.

20 Theodor W. Adorno, 'On Subject and Object', in *Critical Models: Interventions and Catchwords* (Columbia University Press, 2005), 250.

21 Adorno, *Negative Dialectics*, 196; Deborah Cook, *Adorno on Nature* (Routledge, 2014), 11.

object cannot be posited; it does not authorize 'a foundational concept of nature, matter, or the objective world'.[22]

Adorno relates objectivity to the two meanings of materialism, the weight of natural and of social relations on the individual. Both first (unmediated) and second (mediated) nature oppose the subject as objective. In the late Capitalocene, however, there is no longer any unmediated nature. All natural objects, even those that appear unchanging, such as rocks and sediments, or immediate and transhistorical, such as energy, are mediated by social relations (and, ultimately, by the law of value). Mediation first by second nature does not mean that the second constitutes the first – society cannot control or create its own conditions of objectivity. Rather, it means that, throughout history, society and nature have been more deeply mediated through each other, transforming the planet into something that would not have evolved from nature alone. In so doing, society has profoundly transformed itself and the human experience of nature. The asymmetrical relationship of how object and subject are mediated through one another, and the totality of second nature in late capitalism affect the subject's possibilities for knowledge and emancipatory practice.

The difficulties presented by defamiliarizing the experience of energy are the result of its naturalization as a second nature. Machinery first engenders the identification and homogenization of the natural things that compose and are consumed in it. Moreover, it is precisely through technological change that an ever-greater number of activities are drawn into the energy economy: from the sphere of industrial production to the household, from the mechanization to the electrification and digitization of tasks. Technological change is ubiquitous: Within a lifetime, people have moved from heating by coal to gas to the heat pump. Yet across the changes in devices, techniques, engines, and equipment, nature's input has remained apparently the same. Energy, therefore, takes on a certain natural form of appearance. As Alison Stone has put it, within second nature, it seems 'that first nature must prefigure second in

22 Cook, *Adorno on Nature*, 13.

character'.[23] Through the forms of our technically mediated life, nature appears as energetically pre-ordered.

Even the critique of capitalism is delivered in the language of energy. By way of an abstraction and homogenization of production, capitalism has generated, along with a language of value, a second, equally abstract language of matter and energy rivalling monetary accounting and profit. The scramble for profit has historically driven the energetic measurement of the world, and it still does. It is therefore unsurprising that both the fossil and green variants of capital faction draw on this concept in their operations. But the concept of energy has also long been invoked by those *opposing* capitalism. When capitalism's apologists claim that it is the most efficient and productive system, engineers, technocrats, and ecological economists respond that it is, in fact, materially and energetically inefficient.[24] The concept of energy thus emphasizes the material irrationality of production within capitalism. Energy figures both as a category of exploitation and a category of critique.

The Frankfurt School's concept of second nature offers a way to inquire into the ossified forms of second nature, such as energy. Its method focuses on the mediation of what *appears* self-evident, given, and naturally independent. Yet to call 'energy' an 'appearance' is misleading, as the term carries a strong visual connotation. It is true to the extent that *something appears as natural when it is historical* – when its historical character is not recognized or perceived. Yet it is wrong to the extent that its character as nature is not merely a problem of visible form and perspective. Second nature is not something one could alter merely by changing one's perspective, by lifting the veil, or by deconstructing it intellectually. As second nature, energy *is* objective; it exists independently of our thought. However we choose to conceptualize the energy economy today, our lives remain mediated by it. Even the sharpest anti-fossil

23 Alison Stone, 'Adorno and the Disenchantment of Nature', *Philosophy and Social Criticism* 32, no. 2 (1 March 2006): 238.

24 Thorstein Veblen, *The Engineers and the Price System* (Augustus M. Kelly, 1975); Donald R. Stabile, 'Veblen and the Political Economy of the Engineer: The Radical Thinker and Engineering Leaders Came to Technocratic Ideas at the Same Time', *American Journal of Economics and Sociology* 45, no. 1 (1986): 41–52.

manifesto is reliant on fossil fuel. Still, energy is distinct from those areas of the earth untouched by human hands. It is more like the domesticated nature of livestock animals or canalized rivers, a nature fashioned by human labour.

This materialism differs from that of environmental history found in Sieferle or Wrigley. For Adorno, as for Marx, the denial of society's natural prerequisites is an idealist fallacy. On this point, critical materialism agrees with environmental history. Yet the reduction of society's natural condition to a system of concepts would be no less idealist. 'Matter', 'nature', 'energy': None represents 'an immediate given, a point at which theoretical inquiry simply had to stop'.[25] Because nature itself is historical, because it interacts with society and is mediated by a knowing subject, no method, discipline, or conceptual apparatus can guarantee knowledge of nature from the outset. 'The concept of a law of nature is unthinkable without men's endeavours to master nature.'[26] In critical materialism, matter thus appears not as a foundation or ground but as resistance to the subject's practice and as non-identity with its concepts. Adorno warns of a materialism that mirrors idealism's claim to absolute knowledge of the whole.[27] Veinberg's Planet-Generator image expresses this idealist tendency in energetic theory.

Critical materialism seeks fidelity to the object in its search for what resists subsumption by the concept. In a world where the natural is social, and the social all too natural, critical materialism takes on a double task when it confronts the forms of appearance of both nature and society. Adorno formulated this as an attempt 'to grasp historic being in its utmost definition, in the place where it is the most historic, as natural being, or to grasp nature, in the place where it seems most deeply, inertly natural, as a historical being'.[28] Rather than two different realms of being, the historical and the natural are critical categories that

25 Simon Jarvis, *Adorno: A Critical Introduction* (Psychology Press, 1998), 16.

26 Schmidt, *The Concept of Nature in Marx*, 70.

27 Adorno, *Negative Dialectics*, 181.

28 Ibid., 359; see, for a slightly different translation, Theodor W. Adorno, 'The Idea of Natural History', in Robert Hullot-Kentor, ed., *Things Beyond Resemblance: Collected Essays on Theodor W. Adorno* (Columbia University Press, 2006), 260.

can illuminate each other. Articulating the concept of energy from this vantage means abandoning the assumption that what is termed 'energy' today can be subsumed under the abstraction of energy without particularities. It entails a focus on particularities as they escape capture by the generic concept.

The affinity between historical materialism and energy economy discourse must be understood critically if the concepts of the former are to guide us to the non-identical masked by the latter.[29] Both are imbued by the concepts of labour, force, and power, handed down from the eighteenth century, in which nineteenth-century changes in production were experienced and debated in Western political economy and natural philosophy. The capacity of human beings to negate nature through work and bring about new ways of social life – to imagine and realize them – and the changeability, a creative form-generating capacity, of nature did not mobilize the same words arbitrarily. Evidence of this parallel is found, for example, in the long-standing discussion of whether historical materialism does, or should, reflect the discovery of thermodynamics.[30] Some have argued that Marx's concept of labour resembles nineteenth-century engineers' concept of work; others have interpreted the 'metabolism' of nature and society ecologically: as the exchange of matter and energy.[31] Indeed, Marx himself sometimes used language evoking physiology or energy.

The present book does not offer a competing *positive* system; it will not advance a proposed true order of nature to replace energetic

29 Cook, *Adorno on Nature*, 39.

30 J. Martínez Alier and J. M. Naredo, 'A Marxist Precursor of Energy Economics: Podolinsky', *Journal of Peasant Studies* 9, no. 2 (1 January 1982): 207–24; Paul Burkett and John Bellamy Foster, 'Metabolism, Energy, and Entropy in Marx's Critique of Political Economy: Beyond the Podolinsky Myth', *Theory and Society* 35, no. 1 (1 February 2006): 109–56; Thomas Gehring, 'Der entropische Marx: Eine Bitte an den Marxismus, die Entropie-Kirche im thermodynamischen Dorf zu lassen', *PROKLA: Zeitschrift für kritische Sozialwissenschaft* 41, no. 165 (1 December 2011).

31 François Vatin, *Le travail, économie et physique*, 1st ed. (PUF, 1993); Amy E. Wendling, *Karl Marx on Technology and Alienation* (Palgrave Macmillan, 2009); Anson Rabinbach, *The Human Motor: Energy, Fatigue, and the Origins of Modernity* (Basic Books, 1990), 69–83.

concepts. It follows Adorno's negative dialectics in refusing to conceive of nature as the foundation or basic structure underlying society. As the object preponderates, the subject cannot craft a map that would guide it through and give it complete control over its own objective conditions. On the one hand, such maps have a tendency to naturalize what is social: The understanding of nature as structured energetically obscures the fact of energy as the quantification of work embedded in social relations and their reproduction. Capitalist society does not employ nature's work *because* it is inherent in nature as possibility or hidden purpose. Rather, by doing so, it negates nature's spontaneous development. The energy economy can only take forms that are possible in nature, but this potential is not the *cause* of its social constitution; it is the condition of its possibility of achieving objectivity. On the other hand, this naturalization of a social order eclipses the possibilities nature offers to realize vastly different social orders. As Christoph Menke suggests, to think of human freedom within nature, one needs to think of nature not as order but as disorder or negativity: as containing a force that is capable of interrupting itself, dispensing with order, and generating an indeterminate space where cultural forms can take shape.[32]

At the level of nature having entered into the sphere of production, I take up Marx's largely neglected concept of 'natural forces' (*Naturkräfte*), which refers to the manifold ways in which nature changes and takes on new forms. I suggest that it is no coincidence that this concept, which denotes nature beyond human control, has reappeared in the Anthropocene discourse. Marx employs this concept for human beings, who, in their work, confront 'the materials of nature as a force of nature [*Naturmacht*]'. Man labours by using the 'natural forces [*Naturkräfte*] which belong to his own body, his arms, legs, head and hands, in order to appropriate the materials of nature in a form adapted to his own needs'. The nature the worker faces is not only 'slumbering potential' but also 'play of forces', a joyful playfulness that is manifested in the worker

32 Menke, 'Die Lücke in der Natur: Die Lehre der Anthropologie'; for an analysis of how societies have oriented themselves towards natural orders, see Lorraine Daston, *Against Nature* (MIT Press, 2019).

herself, when she 'enjoys [work] as the free play of [her] physical and mental powers [*Kräfte*]'.[33] In her efforts to shape natural materials into things that she can use, the worker is 'constantly helped by natural forces'.[34] Under capitalism, however, this labour process is not under the control of the worker, and all powers of labour (*Kräfte der Arbeit*) project themselves as powers of capital (*Kräfte des Kapitals*). Employed by capital, they bring about capital's forms.[35] The capitalist receives the 'natural force' of labour (coordination, division of labour, etc.) and other properties of nature for free and puts them to work in large-scale production for the accumulation of surplus value.

As should be clear from this summary, the concept cannot be reduced to an energy potential, to the application of a quantity of work. Although motive forces that exist in nature, such as water or wind, may be considered natural forces, so can the soil, or 'the ability of water to change its aggregate state and transform itself into steam . . . the elasticity of steam'.[36] Note that the steam's 'force' consists not in the quantity of its energy content according to Marx, but in its ability to change aggregate states – a qualitative property. Forces cannot be reduced to a single denominator, as they denote nature's creative, playful, form-generating ability – productivity in a rich sense that cannot be quantified. Just as concrete human labour becomes abstract labour for capital, natural forces become sources of energy when nature's manifold mutability is negated and preserved only as a quantity of work. An amount of energy can replace an amount of human labour power, but human labour can never replace natural forces.[37] This is the sense in which I speak of natural forces throughout the book.

Historical inquiry often discovers particularities. There is no common, generalizable manner in which natural things fail to abide by the commodified form of a working nature. Much recent scholarship has

33 Marx, *Capital*, vol. 1, 284–5.

34 Ibid., 134.

35 Ibid., 756.

36 Karl Marx, *Capital: A Critique of Political Economy*, vol. 3 (Penguin, 1981), 782.

37 Schmidt, *The Concept of Nature in Marx*, 96.

indicated the limit of the maximization of work thermodynamics sets in train, and the notion of 'fatigue' developed by physiology to describe the human response to treatment as mere labour power. Both limits can and were expressed as energetic concepts.[38] In my historical study, what resists the appropriation of nature's work can barely be subsumed under a common concept. It is differentiated, whether discussed in terms of minerals (sulphur in coal); the properties of atoms and particles (the drift of electrons in a wire); or the social practice of using uncommodified fuels, such as firewood.

The Book's Structure

The concept of energy is specific to time and place; it does not stem from ahistorical human needs but originates in social praxis. Those social relations giving rise to an abstraction and to the objectification of nature's work are linked to machine production. Following Marx, I understand machine production to mean the historical moment when production is fashioned according to capital's needs, as denoted by the concept of the real subsumption of labour under capital.[39]

This is broadly in line with the insight by historians of science that machines and conversion processes framed the development of the concepts of work, energy conservation, and entropy.[40] Yet I do not wish

38 Rabinbach, *The Human Motor*.

39 Marx, *Capital*, vol. 1, 508; Patrick Murray, 'The Social and Material Transformation of Production by Capital: Formal and Real Subsumption in *Capital*, Volume 1', in R. Bellofiore and N. Taylor, eds, *The Constitution of Capital: Essays on Volume 1 of Marx's 'Capital'* (Springer, 2004), 308–14.

40 Thomas S. Kuhn, 'Energy Conservation as an Example of Simultaneous Discovery', in Marshall Clagett, *Critical Problems in the History of Science* (University of Wisconsin Press, 1959), 321–56; D. S. L. Cardwell, 'Some Factors in the Early Development of the Concepts of Power, Work and Energy', *British Journal for the History of Science* 3, no. 3 (June 1967): 209; D. S. L. Cardwell, 'Power Technologies and the Advance of Science, 1700–1825', *Technology and Culture: Detroit, Mich.* 6, no. 2 (Spring 1965); D. S. L. Cardwell, *From Watt to Clausius: The Rise of Thermodynamics in the Early Industrial Age* (Heinemann Educational, 1971), 188–207; Crosbie Smith, *The Science of Energy: A Cultural History of Energy Physics in Victorian Britain*

to add to the large and comprehensive historical scholarship on the concept of energy in physics. My argument does not hold that all energy-related physics, then and now, flows from the sphere of capitalist production or circulation; it is also not an extension, or historical adumbration, of Alfred Sohn-Rethel's conjecture that all forms of knowledge can be deduced from the law of value. Rather, the present book is more narrowly concerned with the concept of energy *within* the productive sphere of industrial capitalism, where energy is not studied, but produced and consumed, and conceptualized only with regard to its relevance in production and consumption. Unlike James Joule and Hermann von Helmholtz, I follow lesser-known intellectual workers employed by industry and institutions of government-sponsored science. This book can best be understood as a historical-materialist epistemology; it studies how fuel scientists, electrotechnical engineers, and energy economists struggled in their scientific practice to make natural things perform regular, comparable work.[41] It does not advance any strong claims about the concept of energy as applied to other disciplines.

Energy has grown so naturalized, its language taken for granted to such an extent, that its historicization requires some justification. To this end, chapter 1 covers the moment when the concept of energy entered philosophy in the nineteenth century. Energy was not discovered in the manner of a new species or uninhabited island. Rather, it emerged from within a novel perspective, a new ordering principle for relating empirical findings across the fields of mechanics and the study of heat and electricity. A moment of historical consciousness thus accompanied the

---

(University of Chicago Press, 1998); M. Norton Wise, 'Mediating Machines', *Science in Context* 2, no. 1 (1988): 77–113; Vatin, *Le travail, économie et physique*; François Vatin, 'Le "travail physique" comme valeur mécanique (XVIIIe–XIXe siècles)', *Cahiers d'histoire: Revue d'histoire critique* 110 (1 October 2009): 117–35.

41 Hans-Jörg Rheinberger, *On Historicizing Epistemology: An Essay* (Stanford University Press, 2010); Ute Tellmann, 'Kommentar: The Economic or the Economy? – Reflections on the Objects of Historical Epistemology', *Berichte zur Wissenschaftsgeschichte* 37, no. 2 (1 June 2014): 165–9; Thomas A. Stapleford, 'Historical Epistemology and the History of Economics: Views Through the Lens of Practice', in Till Düppe and Harro Maas, eds, *Research in the History of Economic Thought and Methodology*, vol. 35A (Emerald, 2018), 113–45.

rise of the energetic vantage point. Where did this change of view originate? Ernst Mach held that this new perspective was ultimately a product of human evolution, and that changing modes of cognition and sensation engendered the experience of energy. For others, like Ernst Cassirer or Gaston Bachelard, it was the result of a history of the mind: Increasing mathematical formalization and technification altered the manner in which the knowing subject shaped, prepared, and experienced an object. As one of the few historical materialists to contribute to this question, Aleksandr Bogdanov argued that the possibility of experiencing energy was to be understood within a larger social practice, the history of labour, and in particular the rise of machine production. Bogdanov conceived of energy as a new form of causality through which a proletarian world might be built. He thus identified energy with socialist emancipation.

What might Bogdanov's project of a social-pragmatic understanding of energy mean as a form of negative dialectics – a dialectics without resolution? Chapter 2 answers this by sketching the place of energy in contemporary historical materialism. Scholars have long reflected on the 'destructive' moment of energy technology and the contradictions it presents for capitalism. The fossil-capitalist thesis argues that capitalism and fossil fuels have been linked inextricably, that accumulation has become structurally dependent on drawing upon natural materials whose reproduction it cannot control, and whose combustion undermines the reproduction of complex life on Earth in a resulting changing climate. This thesis has moved historical materialism beyond an 'ecological' history to one of energy as a 'social relation' (Matt Huber). Resources are never sought solely for energetic properties; rather, certain properties are prized under capitalist social relations. To capture this affinity between fossil fuels and capitalism, Andreas Malm describes fossil fuels as having an 'abstract' nature. But what is an abstraction of capital in relation to the abstraction of energy? Do not the same capitalist social relations also determine nature's *form of appearance*? I argue that energetic abstraction emerges alongside the real subsumption of the labour process: objectification of the prime mover, machine production of energy commodities, and energetic valuation of production. With

Adorno and Christopher Arthur, I emphasize the contradiction remaining between nature's forms and the capitalist cognition of production; nature is never abstract. Negative dialectics captures traces of the non-identical in their appearance at every stage of homogenization in the energy economy's historical development.

This book begins its historical analysis with the coal economy. Chapter 3 traces coal's resistance to energetic valuation over the long nineteenth century. Calling coal a 'stock of energy', as ecological histories often do, implies a homogeneity and stability of coal's capacity to perform work that does not exist by nature but is created socially and remains precarious. The chapter reveals the transformation of coal from a soft rock yielding fire into quantifiable fuel and raw material for synthetic fuel. In the early steam economy, fuel commodities were quantified and divided according to their weight and those aesthetic qualities of the fire they generated. Until well into the nineteenth century, coal was used without much knowledge of what constituted its capacity to perform work, or which precise properties affected its steam-generating effect. Only when the industrial use of coal in steam engines grew more widespread – in a period when coal travelled long distances via railways and ships, was priced by weight, when its local supply diversified – was the coal material experimentally separated into combustible and non-combustible elements. A quantified concept of fuel was now abstracted from an older aesthetics of fire and focused on the process of combustion and heat. Combustibility was then retrospectively projected onto natural material: What was a relation between nature and society, coal, boiler, and steam engine, then appeared as a combustible substance within coal.

Chapter 4 analyses the means by which electricity developed from a lighting technology based on steam power into the most object-like expression of 'working nature'. Because of electricity's specific materiality, a control of currents could be realized only within closed technical systems where calibration of supply and demand at each moment was a technical necessity. Once realized, however, a homogeneous electric current could be reproduced under the same technical conditions, generated from more sources and transmitted further than steam.

Between 1880 and 1930, the systematic character of electricity at first hampered the expansion of electrical systems, since it led to a technically stable but economically precarious dynamic. What can be seen as a temporal restriction to capital exploitation – electricity cannot be stored – became the basis for the corporate strategies of expansion and intensification. Around the turn of the century, large power companies developed systems enabling the utilization of capital by the moment. In so doing, they converted electricity into an energy commodity whose form was ever more determined by its social function. Electricity became the only type of energy purchased as work per unit of time. The chapter shows how states facilitated these strategies of expansion under the notion of efficiency, thus realizing systems of working nature covering entire regions and national territories. Such systems grew to be so comprehensive that they sparked controversies about the collective management of electric currents of which everyone had become – in principle – a riparian owner.

Today, the energy economy is represented no longer by the grid but rather by the accounting table and the flow chart. In the second half of the twentieth century, energy took on its reified form of an accounting fact. Chapter 5 documents the state and corporate turn to accounting models – that is, energy balances – as the preferred means for governing commodified energy and guiding investment during what has been termed the 'Great Acceleration'. Energy balances set a single unit of energy across the economy. Abstracting from the variety and complexity of energy conversion processes in the gas, coal, oil, and electricity sectors, they assumed that each distinct unit of energy contributes equivalent economic value. Although this method has its roots in the thermodynamic description of engines, governments and industry began to use energy balancing to plan long-term investments in the energy sector. International institutionalization and standardization during the historic restructuring of the global economy of the 1970s signified a dependence on commodified energy and indicated some of the difficulties presented in calculating a balance that might adequately capture global differences in energy use. By formalizing the relationship between energies in the economy, balancing also provided a framework in which energy futures could be imagined; this was crucial for redirecting investment into or away from oil and attempts at decarbonization.

In a period defined by a common understanding of the imperative for radical change in energy use, reflections on the relation between energy and history are particularly germane. The concluding chapter attempts this by returning to Bogdanov's hope as echoed by Veinberg, and shared by so many intellectuals of the time, that nature's work would aid in human emancipation. This promise, prevalent throughout the history of the energy economy in different ways, has apparently failed, some might argue. What could a reconciliation of nature's resistance and human history mean today?

The Planet-Generator illustrator's name remains unknown, as does their intention behind the drawing. It is quite likely that the image was an attempt to represent Veinberg's energetic theory, but it is more telling in its failure to capture the natural conditions and historical changes of this 'conquest' of power. The aim of a critical materialist history of the energy economy is not to uncover what nature truly is and come up with a more truthful representation of it. The energy economy does not *oppose* nature, in the sense of violating nature's 'true' order. Rather, the argument is that the subsumption of the motor mechanism imposed an energetic order on nature. The purpose of this book is therefore not to advance an abstract critique of this order, as this would mean a forgetting of human dependence on it and an abandonment of the progressive meaning that energetic thought once promised – as an expression of the hope for what might be accomplished using nature's work. The aim is rather to revisit this development and uncover its traces of resistance, as they suggest possible alternative forms of humanity's relation to nature in nature.

1

Does Energy Have History?

Energy appears as a natural fact in the energy economy. One may purchase a box of power, as batteries, just as if it had been harvested and packaged like a crate of vegetables. The price of a utilities bill may change from year to year, subject as it is to world politics and the market, but the amount of kilowatt-hours remains fixed (unless one installs new devices or changes one's behaviour). Experts tell us quite precisely how much energy must be replaced by electricity when phasing out fossil fuels; it is a number that is evidently easier to estimate, pertaining more to the realm of an unchanging nature, than are the costs of phasing out fossil fuels. Even in the mostly academic debate on degrowth, fixed material and energetic limits are pitted against the limitless growth of capital. Capital appears abstract and social, while energy appears concrete and material, as a natural property of economic life.

Yet even a superficial glance at the matter casts doubts on the supposed naturalness of energy. Human beings have no *natural sense* for energy. Those substances or processes we call energy today are far too distinct from one another to be experienced as a single phenomenological category (compare the immediate sense of a waterfall to that of solar radiation, and both to a lump of coal). While the quantified measure of temperature still relates to the sensual experience of warmth and cold, the measure of energy bears almost no immediate relation to the senses. Part of the concept's allure stems from how it mismatches and challenges basic experience: The smallest nuclear particles, invisible even to the technically augmented human eye, yield the greatest energy; the immovable features of a landscape – its soil, mountains, and lakes – are propulsive fuel.

Energy, likewise, cannot be called a *natural human need*. Certain human desires – bodily needs and those pertaining to collective habitats – are common across time and space. Human ingenuity devotes itself to creating warmth to protect against harsh weather and the changing seasons; to the illumination of the night-time; to technologies for travelling and performing beyond the capabilities of one's body alone. Many, perhaps all, cultures have words for heat, movement, and light – phenomena to which the concept of energy now refers. To address these needs, however, human societies do not in fact require a concept for energy; they do not need *a quantified understanding of what remains the same in all these processes*. Even beyond individual bodies and their wants, in the sphere of social production, coal was burned, iron smelted, and grain milled, without a general concept linking these activities. The concept of energy abstracts from the manifold ways in which nature enters production; it generalizes nature's work and projects it onto natural material, which only then appears as energy.

In their study of how modern societies have become so dependent on energy use, the humanities and social sciences are bedevilled by the paradox of treating energy as simultaneously natural and historical. On the one hand, following the natural sciences, humanistic and social scientific inquiry considers energy to be a fundamental property of nature that conditions what occurs in time and space, thereby serving as a material basis of human societies. This energetic condition need not be understood as a crude determinism, as was typical of early evolutionary sociologists and energeticists. In environmental and energy history, energy takes on the weaker form of energy entering social or historical analysis as one environmental factor conditioning the social.[1] The

1 Bathsheba Demuth, 'The Walrus and the Bureaucrat: Energy, Ecology, and Making the State in the Russian and American Arctic, 1870–1950', *American Historical Review* 124, no. 2 (1 April 2019): 483–510; Richard White, *The Organic Machine: The Remaking of the Columbia River* (Hill and Wang, 1995).

upshot is that no human society is built from social powers alone. Modern society, especially, must be reminded that it continuously enlists the forces of nature to its ends, while remaining subject to them.

Yet the formulation of the principle of energy conservation and thermodynamics in the mid-nineteenth century constituted such a profound break with the orthodoxy of mechanics that it sparked historical investigations among physicists and other natural scientists. Since then, few other concepts have attracted as much interest among historians and philosophers of science. The concept of energy was at the heart of the epistemological crisis of science in the late nineteenth century, from which the line of thought connecting neo-Kantians, French epistemologists, and Marxist historians of science extends. All emphasized the comprehensive mediation of the concept: the social – technical, economic, and mathematical – conditions under which the concept of energy appeared no longer were merely a context for the discovery of an otherwise neutral piece of knowledge but, rather, affected themselves the concept's form and content.[2] Today, undisturbed by the marriage of ecology and history in one area of energy studies, energy often appears as an imperialist and productivist ideological concept in another.[3]

A field of research whose object is both a transhistorical material fact and a historically produced ideology is bound to produce contradictions. But energy's ambivalence is rooted in both an objective condition of life and an ideologically fraught understanding of how societies are embedded in nature. Theory should not force an artificial unity by dissolving this tension but must explore this contradiction. The present

2 Ernst Cassirer, *Substance and Function and Einstein's Theory of Relativity* (Dover Publications, 1953), 187–202; Gaston Bachelard, *The New Scientific Spirit* (Beacon Press, 1986); Anton Pannekoek, 'Lenin as Philosopher: A Critical Examination of the Philosophical Basis of Leninism', *Philosophy and Phenomenological Research* 10, no. 1 (September 1949): 86.

3 Philip Mirowski, *More Heat Than Light: Economics as Social Physics, Physics as Nature's Economics* (Cambridge University Press, 1989); Cara New Daggett, *The Birth of Energy: Fossil Fuels, Thermodynamics, and the Politics of Work* (Duke University Press, 2019); On Barak, 'Three Watersheds in the History of Energy', *Comparative Studies of South Asia, Africa and the Middle East* 34, no. 3 (1 January 2014): 440–53.

chapter thus returns to the moment when the concept of energy debuted in the Western philosophical tradition. Energy, in this sense, builds upon older concepts of force; both concepts attempt to capture nature's form-generating capacity. There is, however, a distinction, in that force refers to the ahistorical principle conditioning human practice, whereas energy denotes a relation pertaining to history. In the latter, nature is experienced by a sensual, labouring human being. Energy's objectivity is thus rooted no longer, as force was, in the transcendental mind or the divine, but in earthly matters. It is based upon the practice of science, engineering, and industry. Such foundations proved to be consequential at the register of ideas, especially as related to the social imaginary.

Energy Enters Philosophy

In the aftermath of the failed revolution of 1848, a new period of sobriety overtook continental European society. Referring to Georg Friedrich Hegel and Friedrich Schelling, Hermann von Helmholtz observed in 1862 that the 'overly bold Ikarus flight of speculation' had come to an end.[4] Rather than leaping speculation, scientific and social progress would be based on gradual, diligent empiricism. An empirical approach had long animated manufacturing and industries in Britain and elsewhere. Now, France and Germany, joined by other European countries and the US, had embarked on catch-up industrialization that jeopardized British dominance. The decisive shift to steam and coal had occurred even in those regions of France and the US where water power had long dominated the industry (it reappeared as hydroelectricity a few decades later). European empires competed through steam power across the world's oceans as their fossil fuel–driven and coal-forged railways and ships forced open colonial markets.[5] In short, this was a world that had

4 Hermann von Helmholtz, 'Ueber das Verhältniss der Naturwissenschaften zur Gesammtheit der Wissenschaften', Festschrift (1862), in *Vorträge und Reden*, vol. 1 (Vieweg, 1903), 165.

5 Michael Adas, *Machines as the Measure of Men: Science, Technology, and Ideologies of Western Dominance* (Cornell University Press, 2015), 142–3.

already been remade by unleashing and enlisting nature's work, and consciously so. The 'great divergence' manifested itself not only in income but also in per capita energy use.[6] Where industry established itself, the power of fossil-fuelled machinery could gradually be felt and experienced more generally, as railroads, trams, and electricity became more commonly available, first for urban populations, then beyond them. An experience of energetic and technological superiority manifested itself in countless energetic theories of culture.[7]

Amid the successes in putting nature to work, the epistemological status of nature's inherent potential – variously called force, power, or energy – was destabilized. In the thought-world of the eighteenth century, the universe's order had been discerned through the study of regularities in development, and this order was now recast as dynamic. In early-nineteenth-century Britain, political economists turned to the concept of work as a means of denoting the productivity unleashed by machines and the division of labour. As Crosbie Smith and M. Norton Wise have shown, Scottish energy scientists drew upon this concept in their study of motors and steam engines.[8] Likewise, *travail* took hold among post-revolutionary French engineers as well.[9] Moreover, the popularity of the concept of *énergie* reflected the experience that the globe, which had long symbolized stability, 'is now becoming a place of gushing unrest and change'. Political revolutions, then, no longer were

6 Paolo Malanima, 'The Limiting Factor: Energy, Growth, and Divergence, 1820–1913', *Economic History Review* 73, no. 2 (2020): 486–512.

7 Adas, *Machines as the Measure of Men*, 210–21; Eugene A. Rosa and Gary E. Machlis, 'Energetic Theories of Society: An Evaluative Review', *Sociological Inquiry* 53, no. 2–3 (1 April 1983): 152–78.

8 M. Norton Wise and Crosbie Smith, 'Work and Waste: Political Economy and Natural Philosophy in Nineteenth Century Britain (I)', *History of Science* 27, no. 3 (1 September 1989): 263–301; M. Norton Wise and Crosbie Smith, 'Work and Waste: Political Economy and Natural Philosophy in Nineteenth Century Britain (II)', *History of Science* 27, no. 4 (1 December 1989): 391–449; M. Norton Wise and Crosbie Smith, 'Work and Waste: Political Economy and Natural Philosophy in Nineteenth Century Britain (III)', *History of Science* 28, no. 3 (1 September 1990): 221–61.

9 François Vatin, *Le travail, économie et physique*, 1st ed. (PUF, 1993); see also Mirowski, *More Heat Than Light*, 125.

absurd, unnatural outbreaks of violence but could now be seen as a part of the dynamics of nature and human history.[10]

Accelerating political and economic change cohered in such concepts, but they first took the form of principles, rather than a regularity derived from empirical observation. In the eighteenth century, Kantian philosophy had regarded the concept of force, as formulated in the laws of mechanics, as an *a priori*, a principle derived from rational thought alone. Its objectivity was due precisely to its not being an empirical fact. Principles were held to be prior to experience; they conditioned all empirical knowledge (*a posteriori*). Friedrich Schelling and Georg Friedrich Hegel opposed Kant by grasping for a unity beyond the rational *a priori* and the empirical *a posteriori* through speculation: In Schelling, the concepts of force and productivity became principles governing natural and social becoming, or development.[11] With the rise of neo-Kantian empiricism against speculative natural philosophy, the category of force changed its meaning once more: It took on objective connotations precisely because it could be realized, experienced, and measured. Over the nineteenth century, the conceptual field of work, force, and energy thus transformed from referring to a non-empirical principle of changeability to denotations of measurable change.[12]

Faced with the new ambitions of empiricism, European scientific and philosophical traditions underwent an epistemological crisis during the second half of the nineteenth century; the concept of

10 Michel Delon, *Eine Epoche im Umbruch: Die Idee der Energie in der französischen Spätaufklärung (1770–1820). Aus dem Französischen von Heinz Thoma* (Brill Fink, 2022), xvii.

11 Marie-Luise Heuser-Kessler, *Die Produktivität der Natur: Schellings Naturphilosophie und das neue Paradigma der Selbstorganisation in den Naturwissenschaften*, Erfahrung und Denken, vol. 69 (Duncker & Humblot, 1986); Naomi Fisher, 'Freedom as Productivity in Schelling's Philosophy of Nature', in G. Anthony Bruno, ed., *Schelling's Philosophy: Freedom, Nature, and Systematicity* (Oxford University Press, 2020), 53–69.

12 Anson Rabinbach traces a similar shift across different European traditions but focuses on the concept of labour and work. See Anson Rabinbach, *The Human Motor: Energy, Fatigue, and the Origins of Modernity* (Basic Books, 1990).

energy was at its heart.[13] Natural philosophers of the eighteenth century had imagined a universe composed of bodies and forces, with the former attracted and repulsed by the latter, oscillating around a state of equilibrium. Well into the nineteenth century, such mechanical explanations were deemed the most rational and set the standard for all other fields of scientific inquiry. Energy physics challenged this view: While the idea of a 'conservation of force' and the methods for its measurement grew out of mechanics, this line of inquiry soon faced the experimental knowledge of phenomena such as heat, electricity, and life, that resisted easy assimilation into a dualistic concept of bodies and forces. Experiments around heat and electricity demonstrated that these phenomena exhibited a then uncategorizable behaviour that could not be segregated into another 'imponderable' realm of being. Rather, their relation to mechanical phenomena required a profound reconceptualization of the field. A universe of oscillation could hardly be reconciled with the second of thermodynamics, which stated that natural processes may have irreversible directions. The universe – and with it, society – appeared to be defined by a process of development or decline. By the second half of the nineteenth century, scientists and philosophers therefore no longer accepted the mechanical conceptual framework.[14]

The shift from a universal mechanical principle to an empirical mode of science can be observed in the thought of Hermann von Helmholtz. A neo-Kantian, he offered the most philosophically coherent account of what was later to become the first law of thermodynamics in his 1847 lecture 'On the Conservation of Force [*Kraft*]'. There, Helmholtz refers approvingly to the distinction made in Kant's *Critique of Pure Reason* between an 'objective law' of the necessary and universal unity of forces

13 Lorraine Daston and Peter Galison, *Objectivity* (Zone Books, 2007), 211–14; Hans-Jörg Rheinberger, *On Historicizing Epistemology: An Essay* (Stanford University Press, 2010), 1.

14 Peter M. Harman, *Energy, Force, and Matter: The Conceptual Development of Nineteenth-Century Physics* (Cambridge University Press, 1982), 37–8; Crosbie Smith, *The Science of Energy: A Cultural History of Energy Physics in Victorian Britain* (University of Chicago Press, 1998), 101–2, 111.

in nature, and the 'empirical rule' formed from experiences.[15] In this understanding, forces are objective *because* they are not tainted by the subjectivities of perception; forces are an absolute truth. As Lorraine Daston and Peter Galison have discussed, Helmholtz and many others reversed their understanding of 'objective' and the 'subjective', due to exposure to the empirical sciences. By the late 1860s, Helmholtz was less confident in the neo-Kantian legacy of energy physics. Did not laws derived from empirical observation also confront the will as an 'objective power'?[16]

The standard of empirical data as the sole acceptable form of objectivity emerged with positivism as an early symptom of epistemological crisis. As traditional categories in physics evaporated with the proliferation of steam engines, physicists themselves turned to historical investigations to understand the deep conceptual shifts then taking place. Beginning in 1871, Ernst Mach launched a series of studies to trace the use of the term 'energy' as it came to mean an 'indestructible something' common across physical processes, 'of which the measure is mechanical work'.[17] Mach found that scientific practice was riddled with accidents and contingencies. The categories natural scientists used to make sense of their empirical findings were therefore often metaphysical or matters of convention. Rather than 'flowing' from the empirical phenomenon, then, scientific concepts were inherited. To Mach, nothing existed beyond empirical sense perception, and the status as an object itself was the result of abstraction.[18]

If this was so, the space of the absolute and the *a priori*, constituting

15 Daston and Galison, *Objectivity*, 214; P. M. Heimann, 'Helmholtz and Kant: The Metaphysical Foundations of "Über die Erhaltung der Kraft"', *Studies in History and Philosophy of Science Part A* 5, no. 3 (November 1974): 205–38.

16 Daston and Galison, *Objectivity*, 214.

17 Ernst Mach, 'On the Principle of the Conservation of Energy', *Monist* 5, no. 1 (1894): 23; Ernst Mach, *Die Geschichte und die Wurzel des Satzes von der Erhaltung der Arbeit: Vortrag, gehalten in der k. Böhmischen Gesellschaft der Wissenschaften am 15. November 1871* (J. G. Calve, 1872); Ernst Mach, *Die Principien der Wärmelehre: Historisch-kritisch entwickelt* (Johann Ambrosius Barth, 1896); Ernst Mach, *Die Mechanik in ihrer Entwicklung: Historisch-kritisch dargestellt* (Brockhaus, 1912).

18 Rheinberger, *On Historicizing Epistemology*, 9.

the traditional field of philosophy, was now shrinking. After the collapse of speculative idealism, and the failure to integrate the *Wissenschaften* on a mechanical basis, philosophy therefore found itself in crisis. When the empirical sciences divided up the world among their disciplines, what would philosophy be left to do?[19] While scientists and philosophers agreed that there was no longer a philosophical basis for grounding all *Wissenschaften*, they disagreed over a prospective new basis rooted in scientific experience, that experience by which knowledge was relativized, conditioned by technical, practical, and social structures. The main task of philosophy then became the distillation of the 'logic of the sciences', already implicit in the empirical sciences themselves.[20] Scientists treated concepts no longer as absolutes but as 'auxiliary'. They were not 'entities with a higher rank than an individual object' but 'historical products of the logical mind struggling with objects', abstract devices for knowing and exploiting nature.[21] Because the structure of matter could not at the time be proven empirically, whereas energy could be measured, many positivists elevated energy as the primary category of experience.

When science is no longer a product of a transhistorical 'subjective reason', then its discoveries are no longer absolute but contingent on certain times, places, and societies. Certain methodological questions therefore arise: How can truth in theory be distinguished from falsehood? Some scientists and philosophers responded by constructing a new positive system out of empiricist critique. Richard Avenarius, Georg Helm, and Wilhelm Ostwald believed that scientific unity had never been nearer, that a new unity would be based on what underlies all 'experience': energy. Energy here was to be understood not as an object in time and space, but as a quantity emerging from relations among phenomenologically different processes and materials. As a quantity, it existed only at this register of experience. It was a category that promised

19 Frederick Beiser, 'Neo-Kantianism', in Michael N. Forster and Kristin Gjesdal, eds, *The Oxford Handbook of German Philosophy in the Nineteenth Century* (Oxford University Press, 2015), 287.

20 Ibid., 284–5.

21 Rheinberger, *On Historicizing Epistemology*, 12.

to make sense of experimental data. Yet it was simultaneously the transhistorical fact of experience that incorporated being – life, consciousness, and experience itself. Energy linked the object of study with the mind studying it; even viewed through the microscope, the detection of an object in the eye, and its recognition in consciousness, could be expressed in energetic terms. As Wilhelm Ostwald, German chemist and Nobel Prize laureate, put it: 'Everything that happens, happens through energy and to energy.' Energy was seen as grounding a new unity of the world, a new monism, better than matter could: 'Matter', wrote Ostwald, only decades before the quantum was discovered, 'is only a mirage which the mind creates to comprehend the workings of energy.'[22]

The self-proclaimed neo-Kantian Richard Avenarius formulated this new methodological approach in *Philosophy as Thinking the World According to the Principle of the Smallest Measure of Force: Prolegomena to a Critique of Pure Experience* (1876). Rather than a category derived from experience (which was always limited and historical), energy here was already guiding scientific method. According to Avenarius, speculative philosophy à la Hegel had thrown up so many 'pseudo-problems ... causing only confusion and dissipation of energy.'[23] The critical moment in the empirical method was when 'experimental data' was 'conceptualized' or 'interpreted' in thought. This conceptualization was to proceed with the 'smallest measure of force.' Accordingly, Avenarius called his method 'empirio-criticism.' For Mach, who had studied the digressions of eighteenth- and nineteenth-century physicists on their way to the concept of energy, this principle could be observed in reality through all historical aberrations: The things of thought were 'subjected to a historical regime of economy which prevents their random development.'[24] Echoing Avenarius, Mach's methodological principle for a historically self-conscious science is that what is given in experience should be

22 Cited and translated in Niles R. Holt, 'Wilhelm Ostwald's "The Bridge"', *British Journal for the History of Science* 10, no. 2 (1977): 146.

23 Richard Avenarius cited in Klaus Christian Köhnke, *The Rise of Neo-Kantianism: German Academic Philosophy Between Idealism and Positivism* (Cambridge University Press, 1991), 254.

24 Rheinberger, *On Historicizing Epistemology*, 9.

conceptualized with 'the least possible expenditure of thought'.[25] Here, an energetic idea appeared within the operations of the mind to discipline it.

The return to this positive system was likewise indicative of a general movement of German philosophy's response to social democracy and various European strands of socialist materialism. As Klaus Christian Köhnke argued in his study of the neo-Kantians, the retreat from empiricism must be understood as a consequence of the social crisis.[26] The attempted assassination of German Emperor William I in 1878 was particularly shocking to the scientific bourgeoisie; intellectuals rallied to defend the social order. Empiricism was henceforth suppressed in the field of practical philosophy, and less successfully in natural philosophy. Hermann von Helmholtz became a dualist. Whereas he accepted energy physics as a revolution in the order of heaven and earth, the same spirit of experimentation in the social sphere – parliamentarism or socialism – was to be proscribed. Helmholtz lamented that the old *a priori*, which held that 'everything that mankind has hitherto regarded worthy of reverence and gratitude' was united in the figure of the kaiser, was now coming under attack.[27]

Just as neo-Kantianism had transformed itself from a critical into a positivist theory, energetics betrayed its claim to radical empiricism and turned from advancing a critique of the mechanical worldview to the task of defining and stabilizing the categories of social order.[28] Ostwald's infamous attempt to explain human history as a history of ever improving efficiency (*Güteverhältnis*) is only the most obvious breach of empirical method as applied to society.[29] Others, such as Ernest Solvay, studied

25 Ernst Mach, *The Science of Mechanics: A Critical and Historical Account of Its Development* (Open Court Publishing Company, 1960), 586.

26 Köhnke, *The Rise of Neo-Kantianism*, 277–80.

27 Hermann von Helmholtz (1878) cited in Köhnke, *The Rise of Neo-Kantianism*, 268.

28 Marco Vianna Franco has pointed out to me in conversation that these energetic thinkers should not be seen as a coherent group. Their politics ranged from energetic technocracy to cooperative production. See, for a brief overview, Marco P. Vianna Franco and Antoine Missemer, *A History of Ecological Economic Thought* (Taylor & Francis, 2022), 59–75.

29 Wilhelm Ostwald, *Energetische Grundlagen der Kulturwissenschaft* (Alfred

society as an entity ultimately defined by energetic laws, and drew from this the normative claim that positive law should express this character as a community of labour.[30] The reversal of Kantian categories was complete: Mind was now subjective, and the empirical was not only objective but also universal. Anson Rabinbach saw in this energetic tradition a 'transcendental materialism', or the universalization of the concept of work.[31]

From the vantage point of the early twentieth century, philosophers of science, in both the French epistemological and the German neo-Kantian traditions, agreed that the form of scientific inquiry had changed. But they resisted the energeticists' claim that a principle of pure being, the basis of universal order, had been discovered. Ernst Cassirer and Gaston Bachelard both emphasized the mediatedness of the energy concept and focused on how 'form' came to imprint itself on 'content' in this new mode of science. For Cassirer, who focused on the 'mathematization' of science, this 'form' was predetermined by measurement and the concept of number.

> The notion that 'energies' can be seen or heard is obviously no less naive than the notion that the 'matter' of theoretical physics can be directly touched and grasped with the hands. What is given us are qualitative differences of sensation: of warm and cold, light and dark, sweet and bitter, but not numerical differences of quantities of work.[32]

Energy already presupposes mathematization and measurement; it denotes a relation between phenomena, a unitary system of reference, yet not an independent thing. According to Cassirer, we would be mistaken if we were to clothe the principle with the language of a thing:

> Even in the form of 'the' thing, the all-inclusive substance, we should create the same dogmatic confusion, that energism charges against

---

Kröner Verlag, 1909); see, for Max Weber's critique, Max Weber, '"Energetic" Theories of Culture', *Mid-American Review of Sociology* 9, no. 2 (1984): 33–58.

30 Ernest Solvay, 'Soziale Energetik und positive Politik', *Annalen der Naturphilosophie* 9, no. 1 (1910): 105–19.

31 Rabinbach, *The Human Motor*, 49.

32 Cassirer, *Substance and Function and Einstein's Theory of Relativity*, 189.

> materialism. To *measure* a perception means to change it into another form of being and to approach it with definite theoretical assumptions of judgement.[33]

The nature of concepts in science, and hence the form objects could take, had changed from concepts expressing an absolute substance to concepts capturing a function that was always relative and conditioned.

Gaston Bachelard too registered a 'new scientific spirit'. Rather than a reflection of pure experience along the path of least conceptual resistance, as the energeticists maintained, science still operated conceptually, but its concepts had migrated into technical instruments. The empirical sciences relied heavily on such instruments, themselves embodiments of theories, to summon the objects that they study.[34] Concept and object – noumena and phenomena – were now folded into one another. Contrary to the monism of energetics, however, Bachelard maintained that there remained a dialectics in scientific work: Application in the world implied a transcendence of the theory that shaped the instrument.

By recovering Kant's *a priori* in the changed form of a 'logical' or 'conceptual' moment within the empirical sciences, Cassirer and Bachelard also pioneered an understanding of energy as a social and historical relation; energy was the objectification of nature rooted in technology and mathematics. However, this relation was seen as internal to science and not extended to society as a whole. Cassirer parcelled the world of meaning into spheres of different 'symbolic forms' of which science was one, while Bachelard never followed the relation between the technical orientation of science and production.[35] In their theories, science and social practice remained separate. The rise of the concept of

33 Ibid., 192.

34 Hans-Jörg Rheinberger, 'Gaston Bachelard and the Notion of "Phénomenotechnique"', *Perspectives on Science* 13, no. 3 (1 September 2005): 313–28; Bachelard, *The New Scientific Spirit*; Cristina Chimisso, 'From Phenomenology to Phenomenotechnique: The Role of Early Twentieth-Century Physics in Gaston Bachelard's Philosophy', *Studies in History and Philosophy of Science Part A* 39, no. 3 (1 September 2008): 384–92.

35 For a critical appraisal, see Dominique Lecourt, *Marxism and Epistemology: Bachelard, Canguilhem, Foucault* (Verso, 2017), 138–9.

energy in science was studied in isolation from how deeply machines, motors, and the practical application of energetic principles had changed the world around them – from factories to the battlefields of World War I. Over the nineteenth century, nature's inherent capacity for change had transformed from being a transhistorical principle enabling and conditioning human action (but inaccessible to it) to a practical relationship between human society and nature, one subject to historical development.

Energy as a Concept of Proletarian Experience

Around the turn of the twentieth century, energy briefly achieved conceptual prominence in European sociology and reformist movements.[36] Western historical materialism of a reformist hue had, however, never rooted itself in an energetic understanding of society. Only Marxists in the Russian Empire applied themselves theoretically to the integration of energy into historical-materialist method.[37] The

36 Katharina Neef, *Die Entstehung der Soziologie aus der Sozialreform: Eine Fachgeschichte* (Campus, 2012); Chris Renwick, 'The Practice of Spencerian Science: Patrick Geddes's Biosocial Program, 1876–1889', *Isis* 100, no. 1 (1 March 2009): 36–57; Andrew M. McKinnon, 'Energy and Society: Herbert Spencer's "Energetic Sociology" of Social Evolution and Beyond', *Journal of Classical Sociology* 10, no. 4 (11 January 2010): 439–55.

37 To a Western public, the most well-known Russian example of these running debates is Sergei Podolinsky's theorization of the labour theory of value as rooted in the laws of energy. Originally published in German and French journals, it was read by Marx and Engels and still has a place in eco-socialist inquiry. Podolinsky treated energy as a substance of value, though he failed to capture the epistemological debate around it. Sergei Podolinsky, 'Human Labour and Unity of Force', *Historical Materialism* 16, no. 1 (1 January 2008): 163–83; J. Martínez Alier and J. M. Naredo, 'A Marxist Precursor of Energy Economics: Podolinsky', *Journal of Peasant Studies* 9, no. 2 (1 January 1982): 207–24; Paul Burkett and John Bellamy Foster, 'Metabolism, Energy, and Entropy in Marx's Critique of Political Economy: Beyond the Podolinsky Myth', *Theory and Society* 35, no. 1 (1 February 2006): 109–56; John Bellamy Foster and Paul Burkett, 'The Podolinsky Myth: An Obituary Introduction to "Human Labour and Unity of Force", by Sergei Podolinsky', *Historical Materialism* 16, no. 1 (1 January 2008): 115–61.

Dutch Marxist and astronomer Anton Pannekoek later offered an explanation for this difference between East and West: The heated debates on 'scientific materialisms' took place among Russian Marxists in their fight against the tsarist state and the church, while their Western co-thinkers sought to distinguish themselves principally from the categories of bourgeois science.[38]

Among the Russian revolutionary intellectuals, there was an entire group that became highly interested in the German positivists and energeticists. Anatoly Lunacharsky had studied with Richard Avenarius in Switzerland and shared an interest in positivism with Aleksandr Malinowski, better known as Aleksandr Bogdanov, whom he befriended in exile in Kaluga.[39] Bogdanov is now best remembered for his work in *proletkult*, the experiments in blood transfusion that eventually killed him, and his opposition to Lenin. Less well known are his attempts to reconcile historical materialism with a theory of knowledge.[40] Through his Marxist reading of Mach, Avenarius, and Ostwald, he became a pioneering thinker on energy's place in historical materialism.

Like many of his Western contemporaries, Bogdanov believed that Marx's materialism was a relic of the past, made oblivious by the rise of the empirical sciences.[41] To salvage it, it had to be revised and updated on the grounds of a new understanding of science and materialism. This materialism could no longer point to hard things and bodies that were

38 Pannekoek, 'Lenin as Philosopher: A Critical Examination of the Philosophical Basis of Leninism', 148–50.

39 James D. White, *Red Hamlet: The Life and Ideas of Alexander Bogdanov*, Historical Materialism, vol. 172 (Brill, 2018), 39.

40 Maja Soboleva, 'Vom Empiriokritizismus zum Empiriomonismus: Aleksander Bogdanovs Rezeption der Epistemologie von Ernst Mach', in Friedrich Stadler, ed., *Ernst Mach: Zu Leben, Werk und Wirkung* (Springer International Publishing, 2019), 87–97; Maria Chehonadskih, *Alexander Bogdanov and the Politics of Knowledge After the October Revolution: Marx, Engels, and Marxisms* (Springer International Publishing, 2023); Maja Soboleva, 'Zwischen Marx und Mach: Alexander Bogdanows "kritischer Positivismus"', in Aleksandr A. Bogdanov, *Glauben und Wissenschaft: Eine Replik auf Lenins 'Materialismus und Empiriokritizismus'*, trans. Wladislaw Hedeler, 1st ed. (Dietz, 2023), 113–34.

41 James White points out that Marx's theory was in Russia often seen as an 'economic doctrine' that lacked a philosophical basis; White, *Red Hamlet*, 34–5.

represented in the models of mechanics, but it had to deal with the new 'experiences' enabled by technical devices – a more subjective and historical category. Bogdanov thus introduced into Russian Marxism the epistemological question about the condition under which something can become an object of knowledge. His conflict with Lenin and Plekhanov was about the nature of the materialism that rooted Marxism: Unsurprisingly, a 'dematerialization' of the world through scientific discovery was highly disturbing for a Marxist philosophy that understood itself to be scientific and materialist.[42]

Lenin accused Bogdanov of smuggling the idealist, relativist worldview of Western scientists into Russian Marxism. But Bogdanov also developed a Marxist critique of Western positivists. While Avenarius, Mach, and Ostwald spoke of the experience of individual subjects – and understood themselves to be continuing the work of Kant – Bogdanov, by contrast, emphasized the social and practical side of experience as conditioned by production.[43] In linking experience to the totality of social practice, Bogdanov even went beyond the emerging field of social epistemology, which understood the production of knowledge as a collective effort of scientists (see, for instance, Ludwik Fleck's *Denkkollektiv*). He can be seen as pioneering the Marxist approach to the history of science that became so famous after Nikolai Bukharin's and Boris Hessen's contribution to the 1931 London Congress on the History of Science.[44] For Bogdanov, the production of knowledge was no passive registration of sensations, but an active process of working with nature. Indeed, to him, the 'general deficiency' of the empirio-critics was that they continued the problem of the old materialisms, which put forward a 'passive point of view in understanding the world'.[45]

42 Robert C. Williams, *The Other Bolsheviks: Lenin and His Critics, 1904–1914* (Indiana University Press, 1986), 29.

43 McKenzie Wark, *Molecular Red: Theory for the Anthropocene* (Verso, 2016), 21; Aleksandr Bogdanov, *The Philosophy of Living Experience: Popular Outlines* (Brill, 2016), xix–xx.

44 Gerardo Ienna and Giulia Rispoli, 'The 1931 London Congress: The Rise of British Marxism and the Interdependencies of Society, Nature and Technology', *HoST: Journal of History of Science and Technology* 15, no. 1 (1 June 2021): 112.

45 Bogdanov, *The Philosophy of Living Experience*, 150.

The categories through which societies made sense of nature developed within the labour process and also shaped it – as indicated by the category of energy. Bogdanov thus rejected the conceptualization of energy as a neo-Kantian thing-in-itself, beyond human experience, and energy as a mere idea. Both presented energy as detached from the activity of human labour:

> If energy constitutes the basis of all things . . . then it exists in nature completely independently of humanity and its labour activity. If energy is only a symbol, then it exists in human thinking independently of nature – the world of resistance with which society struggles in its labour. In both cases, energy is understood as *cut off* from either of two sides of one objectively indivisible reality. In both cases, it attains an *absolute* character – to be precise, either as an absolute real thing or as an absolute ideal thing. However, in reality, energy is nothing but a practical relationship of society to nature. To see 'energy' in the processes of nature means to look at them from the point of view of how humanity might exploit them in its labour.[46]

In Bogdanov's understanding, energy is both objective and historical. It is how a society that reproduces itself through machines experiences nature, or how it organizes its experience of nature.[47] The society in which it makes sense to speak of energy is one that reproduces itself industrially. Here, nature appears as a unity of interchangeable processes. Abstracting from other sensual qualities, those processes come to be seen quantitatively: 'Energy is not light, sound, heat, etc.; it is the quantitative side of these processes, their measurability and commensurability.'[48] To Bogdanov, energy is the *methodological* abstraction that must be made when nature is put to work. Incisively, he remarks that economists reason in this manner when referring to 'abstract labour', a concept that denotes not the 'labour

46 Ibid., 204, emphasis in the original.

47 Ibid., 212.

48 Aleksandr A. Bogdanov, *Poznanie istoricheskoi tochki zreniia* (A. Leifert, 1901), 11.

of a shoemaker, tailor, blacksmith, but the measurability and commensurability of social labour'.[49]

A New Causality of Proletarian World-Making

Although the energetic perspective arose within the technical intelligentsia during a period of capitalist industrialization, Bogdanov stressed that it did not belong to engineers or scientists alone. Indeed, Western positivists tended to misrepresent or deny altogether the practical and technical dimension of their philosophy. In speaking of the 'labour point of view', Bogdanov sought to reclaim this experience as one belonging to all workers, historically and practically. Insofar as machines were the precondition of the experience of energy, and insofar as they were not invented by individuals but rather 'develop from the economic life of the masses', it is the proletariat's labour, congealed in machinery, that has created the conditions for the energetic perspective. Consequently, for Bogdanov, the proletariat was to seize the opportunities it affords. Machine production enabled the remaking of the world by transforming physical, chemical, and electrical forces into one another, just as natural forces are transformed into the mechanical forces of production. This process amounted to a 'systematic and deliberate transformation of forces or, to be more scientific and exact, in the transformation of energy'.[50] Through machines, the practical application of the 'labour point of view', the proletariat could organize the world to the benefit of human beings.

This practical dimension was reflected in Bogdanov's concept of 'labour causality', a new form of cause–effect relationship that he saw developing and which contained an entire project of world-making. Causality was understood to be no longer a transhistorical structure of experience, as it had been for Kant, but a historically variable condition.[51]

49 Ibid.

50 Bogdanov, *The Philosophy of Living Experience*, 202.

51 Soboleva, 'Zwischen Marx und Mach: Alexander Bogdanows "kritischer Positivismus"', 118.

Bogdanov distinguished a feudal 'authoritarian causality' of order and execution, an 'abstract causality' that had dominated market society, and a 'technological labour causality' made possible by industrialization.[52] This 'labour causality' is modelled on production; cause and effect relate to each other just as the product results from raw materials and labour in production, as intentional transformation. As the 'labour point of view' was not merely contemplative, but practical, the 'technological labour causality' also implied a direction in development:

> We have already said that cognition at its highest level is not only the description and explanation of the world process, but also a programme for world development – not only a map of the theatre of humanity's war with the elements, but also a plan of campaign against the elements. Now we see that the idea of 'labour causality' contains a programme and plan for the conquest of the world – to gradually master circumstances and things in such a way as to purposefully obtain some things from other things and by means of some things to overcome other things.[53]

Rather than a fixed scope of possibilities or abstract laws of energy, 'labour causality', writes Kenneth Jensen, implies the forward striving making 'every process in the world . . . the possible source of every other process'. It suggests that human society has reached a stage where it 'enters the cause–effect sequence as its regulator' and creates the laws and principles to which nature must conform.[54]

Bogdanov defines nature in terms of resistance, but it hardly seems to resist the human project. Nature is 'the world of resistance with which society struggles in its labour', a world that even the societies of his utopian novels never fully overcome.[55] Humanity cannot create something out of nothing in nature. Nevertheless, nature appears to be understood as an

52 Bogdanov, *The Philosophy of Living Experience*, 201, 205–6.

53 Ibid., 205.

54 Kenneth Martin Jensen, *Beyond Marx and Mach: Aleksandr Bogdanov's 'Philosophy of Living Experience'* (D. Reidel, 1978), 120–1.

55 Bogdanov, *The Philosophy of Living Experience*, 204.

object of domination. Breaking its resistance is already anticipated. It is evident in Bogdanov's description of energy that resistance serves only as a moment that can and should be overcome.

> To recognize that a certain phenomenon – an ocean tide or a piece of coal, for example – contains a certain sum of energy means to believe that if one succeeded in completely mastering that phenomenon and successfully utilizing it for the goals of labour, *then it could serve to overcome a certain sum of the elemental resistance of nature*. The principle of energy is the ideal of the power of society over nature.[56]

Energy and resistance are interdependent. Although nature's resistance cannot be wished away, its form is already known in advance, and its overcoming is anticipated. The level of resistance corresponds to a quantity of energy. It remains unclear whether the 'labour point of view' can experience nature's resistance only energetically, whether nature is already energetic in form (considering Bogdanov called his philosophy an 'empirio*monism*') or whether the former presupposes the latter. In any case, Bogdanov's theory contains no strong concept of nature's resistance and, thus, of dialectics.[57]

There is a history of energy. For Mach, Cassirer, and Bogdanov, it was never an eternal object but the result of taking on a new 'point of view'. Yet whether energy was an outgrowth of natural or social history remained in dispute. Was its emergence itself ultimately natural, rooted in the biological life of the species that made sensual knowledge possible, as Mach claimed? Or was it a result of the history of forms of the mind or of the scientific spirit, as Cassirer and Bachelard claimed? Does it belong to the realm of freedom or necessity? This question became

56 Ibid., my emphasis.

57 Compare this, for instance, with Andrei Platonov's description of nature's 'hardness' in 'The First Socialist Tragedy' (1936): Here, it is the very resistance of nature (also called dialectics by him) which protects human beings from devouring it entirely and hence saves them (and nature) from self-destruction. Andrei Platonov, 'On the First Socialist Tragedy', *New Left Review* 69 (1 June 2011): 31–2.

particularly pressing once the concept of energy was understood as a practical guide.

For the energeticists Wilhelm Ostwald and Ernest Solvay, the concept of energy revealed a natural order that rooted the character of societies as productive organisms:

> Socially speaking, man has a purpose in life: he is an energetic apparatus compelled to contribute to the realization of general production. Man is energetically productive by virtue of his vital necessity when he is alone, and he becomes so through the obligation of reciprocity when he is a member of a group; from this essential and fundamental obligation of reciprocity flows by necessity the scientific, energetic-productive norm [*Recht*], which must necessarily form the constitutive basis for the future organisation of social groups.[58]

The 'necessity' invoked by Solvay stems from the energetic nature of human beings and societies. The energeticists' answer to the social question was the enforcement of an energetic order, legitimized by its purported insight into nature, by way of ethical and legal norms. Solvay backed a positive politics and legal measures to regulate the energetic reciprocity that existed in society by nature, and Ostwald famously formulated the 'energetic imperative' proscribing wasted energy. Workers should be obliged to contribute to social production, and, in return, would be guaranteed a wage matching their energy expenditure. Though Solvay and Ostwald understood themselves as progressive social reformers, their ideas amounted to little more than the scientific determination of the lower limit of the wage, corresponding to bare subsistence.[59]

Despite Bogdanov's debt to the energeticists, his conceptualization of a society grasping its own energetic condition differed from theirs considerably. For Bogdanov, energy remained linked to history and to

58 Solvay, 'Soziale Energetik und positive Politik', 107.

59 Vatin, *Le travail, économie et physique*, 117; Wilhelm Ostwald, *Der energetische Imperativ* (Akademische Verlagsgesellschaft, 1912).

proletarian emancipation and was open-ended. Bogdanov held that under socialism the development of productive forces might engender progress for the human species as such. Mechanical production enabled and anticipated such an outcome, but it could not by itself realize it. This was no formal problem that would be solved by 'installing' socialism from above; organization was required (hence the need for *proletkult*). Bogdanov took the task of organizing a socialist energy economy seriously; it could not mean copying the principles of the empirio-critics. He criticized them for making energy conservation a general social norm: 'Victory over nature is achieved not by petty conservation of energy, but by its fullest, most productive use.'[60]

Seen in the light of twentieth-century developments, Bogdanov appears to be among the last Marxists whose thought preserves an understanding of the historicity of energy in its connection to production. By this interpretation, Bogdanov understands energy as historical and objective at once. This hermeneutic is no mere approximation of a concept to match a 'true' object but is rather a process of a social practice of comprehension through the experience of it. Energy science, then, is not external to energy's production; it is not an objective authority to which Marxism must be adapted but a field of knowledge to be understood within the totality of social practice.

There is also, in Bogdanov's understanding of history, an unsalvageable positive dialectic. For Bogdanov, the progression of proletarian experience *is* that of overcoming resistance and of the domination of nature. But the failure of this project of energetic, proletarian world-making precludes a naïve return to his ideas, as he wrote in anticipation of socialism, rather than mainly against the prevailing capitalist society.

60 Bogdanov, *The Philosophy of Living Experience*, 147; Jensen, *Beyond Marx and Mach*, 75–80. Bogdanov also opposed the introduction of scientific management (sanctioned by Lenin), arguing that production could only be understood as rationalized once it increased productivity rather than labour intensity. Taylorist methods, however, were merely a means of intensifying exploitation – they would require a greater expenditure of natural and human energy. See Aleksandr Bogdanov, *Mezhdu chelovekom i mashinoi: O sisteme Teilora* (Ukrainskogo Centralnogo Agentstvo po Snabzheniiu i Rasprostraneniiu Pechati, 1919), 10.

Bogdanov does not consider the persistence of certain social antagonisms under really existing socialism, that socialist machines will not be overseen by 'humanity', that some will decide on their use, others will be subject to it, and that the pursuit of energetic world-making might undermine the conditions for life on earth. That this resistance would be as material and chemical as it was energetic was never expressed in his work. But what might it mean to transform Bogdanov's positive project of understanding energy within social practice into a negative dialectics that no longer anticipates reconciliation of nature and society by means of energy?

2

Forces of Nature, Sources of Energy

Humanity has expended a staggering twenty-seven zettajoules of energy since the Industrial Revolution. In the last seventy years alone, it has consumed more energy than in the entire 11,700 years of the Holocene.[1] Although small in cosmic terms, this is the largest amount of energy ever to be captured on the surface of the earth. The planet has never been more energetically 'active'. All changes in the Earth system to which the Anthropocene thesis refers are mediated in some way by this massive quantity of energy, whether they take the form of nitrogen cycles, land use, ocean acidification, or the chemical changes in the atmosphere. Human energy use has left traces in the natural history of the planet: The Anthropocene is an energy-historical event. Humanity – to stay with the language of the Anthropocene for a moment longer – has produced a state of the planet that would not have occurred naturally: It has yielded changes, as it were, both in nature and *against* nature. In the energy economy, humanity appropriates nature's capacity to change and turns it against nature, to change it. But why speak then

1 Jaia Syvitski et al., 'Extraordinary Human Energy Consumption and Resultant Geological Impacts Beginning Around 1950 CE Initiated the Proposed Anthropocene Epoch', *Communications Earth and Environment* 1, no. 1 (16 October 2020): 3.

of humanity as a geophysical *force* rather than, let us say, planet-making *labour*?

Contrary to what Aleksandr Bogdanov envisioned, the global energy economy is not the product of proletarian world-making. There has never been a collective, conscious subject planning and realizing the energy economy as it exists today. The vast majority of humanity never had any say about whether, and for what purpose, nature was put to work, and even those who could control a certain amount of energy never determined the direction and momentum of the entire system. What had occurred to Boris Veinberg as a mastery of planetary forces – the production and consumption of energy, and particularly of fossil fuels – has turned out to be something else: a largely unintentional tinkering with planetary dynamics, producing geological changes that will last many thousands of years. The world built through a control of nature's work has yielded uncontrolled changes in nature and society. Humanity is no working collective but a blind geophysical force.

The re-emergence of the concept of force in the Anthropocene discourse must be properly contextualized. Over the course of the nineteenth century, a language invoking 'forces' or 'agents' of nature gave way to references to sources of power or energy.[2] The transition is gradual and differentiated. There is much discursive ambiguity, as we have seen in the last chapter, where concepts of force, power, and energy are used across natural philosophy, political economy, and the emerging science of energy. There is variation across languages, too. Still, when one compares the bookends of this period – from the seventeenth to the late twentieth centuries – clear differences emerge. 'Energy' was rarely used in reference to the employment of natural forces in production prior to 1800. Today, the few remnants of the concept of force in industry, especially in the Germanic languages (e.g., *Kraftwerk*), evoke previous epochs. This discursive transformation indicates a changing relation to nature on the part of industrializing societies. Indeed, it signifies a more extensive appropriation of natural

2 Cornelia Zumbusch, *Romantische Thermodynamik: Dichtung, Natur und die Verwandlung der Kräfte 1770–1830* (De Gruyter, 2023), 17–18.

forces in which the latter appear to be socially controlled – in a process I call the objectification of energy.

The transition from 'forces of nature' to 'sources of energy' over the last two centuries indicates a break with a worldview in which all nature, inclusive of humanity, had been a site of force. Forces were understood as an active principle to be directed but not perfected. On its own, a force sets objects in motion, shapes matter, or establishes a balance by acting against other forces. Formalized in Newton's physics, then elaborated upon in Leibniz and Descartes, the concept of force became a central part of Enlightenment thinking (it entered Kantian thinking as an *a priori*, as we have seen in the previous chapter). Eighteenth-century natural philosophers were of two minds about the quantity that was ultimately thought to be conserved in the universe. The two fractions corresponded to Leibniz's *vis viva* (living force, or motive force) and Newton's momentum, but each related itself to the concept of force. Although debate between them took place within the framework of a rational mechanics, the concept of force soon exceeded its narrower scientific definition.

Romantic thinking dealt with force as an excessive and immaterial principle immanent in nature. Forces, for the Romantics, referred not to things, but to transformative relations between objects. Force described the structure of one thing affected by another in such a way that the former is transformed by the latter. 'The structure of force', writes Christoph Menke citing Johann Gottfried Herder, 'is "the operation of one thing into another."'[3] Expressive and form-generating, forces pointed beyond and transformed the existing world. A self-reproducing striving permeated the natural and the human world and could be witnessed as much in art as in electric phenomena. From Schelling to Goethe, from Justus Liebig to, indeed, Karl Marx, the notion of *Kraft* mediated between necessity and freedom without identifying those terms with either nature or society. At the same time, political economists found a common activity in nature and society as well: labour, work, or *travail*. If not

3 Christoph Menke, 'Force: Towards an Aesthetic Concept of Life', *MLN* 125, no. 3 (2010): 558–9.

expressive in the sense of art, nature still appeared as active and value-producing; it offered work 'as a productive service of natural agents', as Jean-Baptiste Say put it.[4]

The difference in perspective between natural forces and sources of energy is telling. Those natural forces referred to autonomous, self-governed motion of animate and inanimate agents. In mechanics and economics textbooks of the first decades of the nineteenth century, machines, for instance, were described as receiving their motion from a part of nature moving of its own accord: They do not 'actually produce that force which arises from some natural cause'.[5] The machine therefore *borrows* motion that is a manifestation of a natural agent or force beyond human control. In water power, the mill owner uses the force of gravity. By means of this ingenious trick, human beings employ nature's forces – the principal agents of change – against nature. A machine acts like a 'canal for work'; it streamlines and redirects motion towards an object of human purpose.[6]

Sources, in contrast, imply a passive store of motive power. The steam engine was at first described as employing the force of heat to produce a mechanical effect to be utilized.[7] But how 'autonomous' a force is heat when human beings can produce and contain it technically? When it can be commanded into existence everywhere at will? The language of forces ceases to make sense with the steam engine. Through the mediation of heat, the motor becomes 'self-acting' and power could now simply be sourced from the 'immense reservoir' of nature.[8] Contrary to forces,

4 Cited in François Vatin, *Le travail, économie et physique* (PUF, 1993), 29.

5 John Farey, *A Treatise on the Steam Engine* (Longman, Rees, Orme, Brown, and Green, 1827).

6 Dionysius Lardner, *The Steam Engine, Familiarly Explained and Illustrated; with an Historical Sketch of Its Invention and Progressive Improvement* (E. L. Carey & A. Hart, 1836), 19; Gaspard Coriolis, *Du calcul de l'effet des machines, ou Considérations sur l'emploi des moteurs et sur leur évaluation, pour servir d'introduction à l'étude spéciale des machines* (Carilian-Goeury, 1829), 27.

7 James Renwick, *Treatise on the Steam Engine* (G. & C. & H. Carvill, 1830), 112.

8 Sadi Carnot, *Reflexions on the Motive Power of Fire: A Critical Edition with the Surviving Scientific Manuscripts* (Manchester University Press, 1986), 38.

which realize themselves, 'source' denotes a passive potential that must be tapped to be realized – sources require a complementary active principle. Historians have argued that the idea of the 'resource' took wood as its first referent. This material had been used for millennia for a wide range of purposes and is a product of the relatively controllable domain of the forest.[9] Sources are constituted by a natural deposit that is already ordered according to its utilization: a managed forest, a channelled river, a prepared bulk of coal. Through the conceptualization of source, agency now found expression in technology, rather than nature. Boilers could generate the force of heat, valves and pipes could control it, and the engine could apply it to work. Technology appeared self-acting, while nature's contribution seemed almost passive, a latent feature waiting to be activated. This discursive shift becomes evident in the worldview of electric capital in the twentieth century: The international organization representing this power economy, the World Power Conference (Weltkraftkonferenz), published a booklet on 'power resources of the world' in 1929, presenting a global survey of the 'leading sources of energy, such as hard coal, brown coal, coke, oil, gas, water-power'.[10] With steam power and electricity spreading geographically and across industries, the language of natural forces gave way to a language of energy and its sources. Nature had now become a source of work, ready-made for capital.

It is appropriate that the language of force should reappear in the Anthropocene discourse to lend expression to the experience of an uncontrolled sourcing of power altering the planet. The consequences of energy use were always beyond humanity's imaginative and practical abilities when acting as an excessive, uncontrollable geophysical force, as

9 David F. Lindenfeld, *The Practical Imagination: The German Sciences of State in the Nineteenth Century* (University of Chicago Press, 1997), 29; Paul Warde, *Ecology, Economy and State Formation in Early Modern Germany* (Cambridge University Press, 2006); Andrea Westermann, 'Geology and World Politics: Mineral Resource Appraisals as Tools of Geopolitical Calculation, 1919–1939', *Historical Social Research* 40, no. 2 (2015): 154.

10 Hugh Quigley and D. N. Dunlop, *Power Resources of the World (Potential and Developed)* (World Power Conference, 1929), vii.

if possessed by an unfathomable natural force itself. I interpret the discursive reappearance of force as signalling an experience of excess and loss of control, an experience of how the energy economy turned into its opposite – not the human control of nature's changeability for the reproduction of humanity, but uncontrolled natural change. The purpose of this chapter is to sketch a conceptualization of the energy economy in light of what it has become today. How does one conceptualize the social and natural antagonism of the energetic forces of production? What role does the abstraction of energy play in it? And how can nature's resistance to that abstraction remind us of the lost meaning of natural forces?

The Social and Natural Antagonism of the Energetic Means of Production

In the old meaning of the word, 'technique' refers to man-made objects that imitate nature to help people live from it. Between the subject's hand and the object to be worked on, the tool acts as an extension of the human body and increases the subject's ability to manipulate the object. Energy technology can be seen as a particular type of tool that makes it possible to scale and accelerate, regulate and homogenize the power of a natural force, which is then transmitted to a working instrument. From this isolated perspective, energy history has often been written as a century-long progress in manipulating nature – more sources could be harnessed more efficiently, new forms of energy could be created and controlled, and the application of power could be scaled from giant machinery to the atomic level. Yet technologies such as the textile mill's waterwheel, the maritime steam engine, or the regional electrical grid always already exist in a social context. Their components cannot be treated in isolation, as they are embedded in a larger technical, economic, and social environment. As part of the productive forces, energy technology is conditioned by social relations.

Under capitalism, the development of technology, including energy technology, is not a neutral or autonomous process for increasing human labour capacities. The machine is not an independent category, but a component of an economic regime. The emergence of that part of machinery called *energy technology* is the result of historical development. Only with large-scale machinery – motor, transmission, and instrument – can one speak of a 'prime mover' separate and simultaneously integrated into production. The steam engine did not set off the Industrial Revolution, but it constituted a 'second revolution' enabling the large factory.[11] Later, electricity, which remained primarily steam-bound, became the technical precondition for rationalization and digitalization. Oil undergirded the global circulation of goods. Progress in energy technology is propelled by competing companies and states, rather than humanity as such.[12] A social antagonism exists between those owning and employing energy technology, and those labouring in factories powered by natural forces: Gains in productive power come at the cost of undermining workers' lives. Under capitalism, productive technology is always also destructive.

In a historical-materialist understanding of the development of the productive forces, their destructiveness is a moment within the formation of the proletariat, to be overcome by proletarian appropriation. In *The German Ideology*, Marx and Engels write that under capitalist relations of production, all means of production turn into tools to control and expropriate workers. The forces of production 'only cause mischief, and are no longer productive but destructive forces'. As the proletariat bears 'all the burden of society without enjoying its advantages' it is 'forced into the sharpest contradiction to all other classes'.[13] Subject to all the destructiveness, the proletariat has nothing to lose and everything to gain from revolution. By

11 Andreas Malm, 'Marx on Steam: From the Optimism of Progress to the Pessimism of Power', *Rethinking Marxism* 30, no. 2 (3 April 2018): 176; Amy E. Wendling, *Karl Marx on Technology and Alienation* (Palgrave Macmillan, 2009), 138; Karl Marx, *Capital: A Critique of Political Economy*, vol. 1 (Penguin Classics, 1992), ch. 15.

12 Leo Marx, 'Technology: The Emergence of a Hazardous Concept', *Technology and Culture* 51, no. 3 (2010): 561–77; Robert L. Heilbroner, 'Do Machines Make History?', *Technology and Culture* 8, no. 3 (1967): 335–45.

13 Karl Marx and Friedrich Engels, *The German Ideology* (International Publishers, 1972), 94.

appropriating the machines and freeing them from their limits under private industry, the proletariat could refashion them into universal means of production – destructive for no one.

The social and natural antagonisms of energy technologies created a situation in which the prospect for proletarian reappropriation and a more collective organization could emerge. One contradiction concerns the limits of private exploitation of raw materials once this had become a general condition of production. The scramble to maximize profit could lower productivity in some industries and jeopardize the security of supply to others. The extractive industries, for instance, had always dealt with the fact that the material to be mined was not organized for this purpose, deposits were of unknown size, limited, and subject to declines in quality. Prices thus fluctuated dramatically. The pattern was most evident in petroleum fields, where taps released pressure and thus degraded the conditions of subsequent extraction.[14] Attempts to nationalize coal industries or to manage the electric current in large national systems likewise reflected this conflict between the general interest and the interest of individual producers. The nature and sustainable operation of these industries indicated a tendency towards collective organization.[15]

14 The creation of monopolistic or cooperative bodies to manage oil production for conservation purposes (i.e., to maximize value) can be traced from Standard Oil's control of the railways and refineries, to the Texas Railroad Commission and the Organization of the Petroleum Exporting Countries (OPEC); William R. Childs, 'Origins of the Texas Railroad Commission's Power to Control Production of Petroleum: Regulatory Strategies in the 1920s', *Journal of Policy History* 2, no. 4 (1990): 353–87; Giuliano Garavini, *The Rise and Fall of OPEC in the Twentieth Century* (Oxford University Press, 2020), 64, 95; Michael Dobson and Giuliano Garavini, 'Juan Pablo Pérez Alfonzo and the Invention of Anticolonial Democratic Oil Conservation', in Daniela Russ and Thomas Turnbull, eds, *Energy's History: Toward a Global Canon* (Stanford University Press, 2025), 129.

15 Carsten Burhop and Thomas Lübbers, 'Cartels, Managerial Incentives, and Productive Efficiency in German Coal Mining, 1881–1913', *Journal of Economic History* 69, no. 2 (2009): 500–27; John E. Murray and Javier Silvestre, 'Integration in European Coal Markets, 1833–1913', *Economic History Review* 73, no. 3 (2020): 679–80; Adam Plaiss, 'From Natural Monopoly to Public Utility: Technological Determinism and the Political Economy of Infrastructure in Progressive-Era America', *Technology and Culture* 57, no. 4 (2016): 807.

As Timothy Mitchell has argued, the general dependence on coal and the large network required to keep coal flowing in industrialized countries allowed for the organization of workers along the supply chain from mine and dock and factory.[16] Workers were aided in their revolt against the conditions under which they work by the many antagonisms that open between the energy technologies, the resources they consumed, and the relations of production.

The capitalist deployment of energy technology also produced natural antagonisms beyond the worker's body and the productivity of a particular industry. It set off processes that threatened larger swathes of the population and jeopardized reproduction at the level of society – a contradiction discussed today under the concept of 'metabolic rift'. Energy technology was already implicated in this initial formulation of how humanity had become estranged from its natural condition under capitalism. The steam engine intensified urbanization as employment in factories drew more workers into the city. When a small rural population, aided by chemistry, supplied a much larger population in the cities with their agricultural produce, it 'disturbed the metabolic interaction between man and the Earth'.[17] Nutrients from the soil flowed into the cities, where night soil accumulated and caused hygienic problems, while the quality of the agricultural land deteriorated.[18] What large-scale industry running on steam had harmed, electricity could perhaps heal. After the first long-distance transmission of electricity, Friedrich Engels connected the possibilities of electricity to a more equal distribution of industry and workers over space and hence, a 'sublation of the antagonism of the city and countryside'.[19] Before the rise of the petroleum-run internal combustion

16 Timothy Mitchell, *Carbon Democracy: Political Power in the Age of Oil* (Verso, 2011).

17 Marx, *Capital*, vol. 1, 637.

18 Ibid., 636–9; see also John Bellamy Foster, 'Marx's Theory of Metabolic Rift: Classical Foundations for Environmental Sociology', *American Journal of Sociology* 105, no. 2 (1 September 1999): 366–405.

19 Karl Marx and Friedrich Engels, *Marx-Engels-Werke*, vol. 35, *Januar 1881–März 1883* (Dietz, 1979), 444–5.

engines enabled individual motorized transport and suburbanization, electricity was widely seen as having a 'rationalizing' and 'socializing' character that could overcome the contradictions and the destructiveness caused by the first Industrial Revolution.[20]

Put more abstractly, limits to resources plague capitalism. No social form within capitalism can overcome the difference between natural things that reproduce themselves on timescales far beyond those of industry and capital's incessant drive to accumulate, which can only realize value within nature. Importing the 'form' of nature from ecology and thermodynamics, some Marxists have described capitalism as having a double character, a system of value engendering a system of matter and energy flows.[21] Although the two never coincide – value does not directly express an amount of energy or matter – the accumulation of value still develops in terms of matter and energy within a specific capitalist energy system. As part of nature, this system is ruled by a 'thermodynamic' principle of declining quality and availability. Capitalism can adapt, but it can never transcend this limit.[22] Yet this loosely thermodynamic reformulation of the ecological problem under capitalism is both too abstract and too specific. It offers no basis for critique as a *determinate* negation. It is too narrow to capture the diversity of ecological problems – it does not confront toxic run-off, nor can climate change be grasped as a problem of energy degeneration. It is also too abstract in the sense that thermodynamically conscious production – the positive side

20 Matthew T. Huber, *Lifeblood: Oil, Freedom, and the Forces of Capital* (University of Minnesota Press, 2013); Patrick Geddes, 'The Twofold Aspect of the Industrial Age: Paleotechnic and Neotechnic', *Town Planning Review* 3, no. 3 (1912): 176–87; Lewis Mumford, *Technics and Civilization* (University of Chicago Press, 2010).

21 Elmar Altvater, 'The Social and Natural Environment of Fossil Capitalism', *Socialist Register* 43, no. 43 (2007): 37–49; Paul Burkett and John Bellamy Foster, 'Metabolism, Energy, and Entropy in Marx's Critique of Political Economy: Beyond the Podolinsky Myth', *Theory and Society* 35, no. 1 (1 February 2006): 109–56. Similarly, Jean-Claude Debeir, Jean-Paul Deléage, and Daniel Hémery, *In the Servitude of Power: Energy and Civilisation Through the Ages* (Zed Books, 1991).

22 Nicholas Georgescu-Roegen, *The Entropy Law and the Economic Process* (Harvard University Press, 1971).

of the critique – is inadequate for understanding human needs and values. A feast, from this perspective, is a waste of energy and nothing more.[23]

The most recent conceptualization of the antagonism between the relations and energetic forces of production is focused on capitalism's use of fossil fuels. Scholarship on fossil capitalism argues that capitalism reproduces itself only through the deterioration of planetary conditions caused by the combustion of fossil fuels. The thesis allows for various readings, from the historical to the structural. More narrowly it is a thesis about capitalism's adoption of fossil fuels and is part of a historical-materialist explanation of the generalization of fossil fuel use. The thesis can also express a structural relationship between fossil fuels and capitalism: Capitalism turned out to be fossil capitalism because it cannot realize itself other than as a fossil-fuelled mode of production. What began as a mere 'historical alliance' has become a structural necessity that is 'impossible' to overcome within capitalist relations of production.[24] A natural antagonism reappears here. Capitalism relies upon a natural material it cannot reproduce; worse still, its combustion undermines the reproduction of living labour on which it depends.

By studying the uptake of coal within capitalist social relations, scholarship on fossil capitalism also stresses the inability of reducing fossil fuels to their energetic characteristics. Such fuels are not useful qua their energy content. Echoing Bogdanov, Matt Huber argues against the fetishization of fossil fuels 'as a "thing" or a "resource"' and for 'a conception of energy as a "social relation" enmeshed in dense networks of power and socioecological change'. Rather than invoking transhistorical

23 Thomas Gehring, 'Der entropische Marx: Eine Bitte an den Marxismus, die Entropie-Kirche im thermodynamischen Dorf zu lassen', *PROKLA: Zeitschrift für kritische Sozialwissenschaft* 41, no. 165 (1 December 2011): 638–42. To borrow a phrase from Engels, it seems ecologistic eco-Marxism was led astray because it was searching for a natural-scientific proof for the necessity of socialism – which it thought it found in the abstraction of energy. See Marx and Engels, *Marx-Engels-Werke*, vol. 35, 135.

24 Elmar Altvater, 'The Capitalist Energy System and the Crisis of the Global Financial Markets: The Impact on Labour', *Labour, Capital and Society / Travail, capital et société* 40, no. 1/2 (2007): 22; Altvater, 'The Social and Natural Environment of Fossil Capitalism', 45.

energy principles and abstract energetic limits, historical materialism should historicize and contextualize resource use. Such analysis should focus on the conditions under which the properties of fossil fuels came to matter and the social and economic consequences they engendered.[25] Fossil fuels were a catalyst for capitalism; they do not in themselves cause but rather 'hasten the generalization and extension' of a capitalist world economy.[26]

Fossil fuels accomplished this by facilitating the exploitation of labour. In Andreas Malm's analysis of the British textile industry, he demonstrates that there was no general scarcity of water that could have explained millowners' uptake of steam. Contemporary observers of industrialization pointed to a quite different advantage: Steam could be employed where labour was cheap and cooperative. Thus, the mobility of steam engines and coal enabled industrial production to move to cities and to hire a disciplined labour force for a guaranteed period of working time. What gave steam a final edge over water was the introduction of working-day regulation in the 1830s. While water-powered production was bound to the rhythm of the rivers, steam-powered mills could marshal a stable source of power for the entire working day.[27]

Tapping into a mobile stock of energy, steam power achieved an abstraction of time and space in industrial production. It was a temporal and spatial pattern congruent with capital's own worldview.[28] Capital creates an 'abstract space' in its image, 'a matrix of nodes and arteries that evolve not through their revealed biophysical attributes but through the circuit of capital'.[29] Within this abstract spatio-temporal sphere, achieved by this new mobility and flexibility of the energetic means of production, capital may freely exploit differences in labour discipline and cost. Thus,

25 Matthew T. Huber, 'Energizing Historical Materialism: Fossil Fuels, Space and the Capitalist Mode of Production', *Geoforum* 40, no. 1 (1 January 2009): 106; Huber, *Lifeblood*, 3.

26 Huber, 'Energizing Historical Materialism', 110.

27 Andreas Malm, 'The Origins of Fossil Capital: From Water to Steam in the British Cotton Industry', *Historical Materialism* 21, no. 1 (1 January 2013): 41–2.

28 Altvater, 'The Social and Natural Environment of Fossil Capitalism', 41.

29 Andreas Malm, *Fossil Capital: The Rise of Steam Power and the Roots of Global Warming* (Verso, 2016), 301.

coal-powered steam brings the physical reality of production closer to capital's abstract requirements. Malm claims that these advantages of fossil fuel made it indispensable to capitalism from that point on.

What unfolds in this abstract space and time is an abstracted form of labour. In other words, labour, an activity unfolding in a concrete time and space and bound to a human body, is likewise treated independently of its concrete manifestation. Malm has shown the discipline of the faculty of human labour both in his study of the colonies around water-powered mills and in the deeper exploitation of urban workers that steam mills enabled. As I argue here, this abstraction of the labour process in time and space also implies an abstraction of the work performed by natural forces. Flowing water and expanding steam are treated according to their capacity to do equivalent work for the millowner, a direct linkage to human labour's existence under an exploitative regime of production. The activity of non-human nature and human beings is abstracted, their performed work can be substituted, and the production process approximates a pure motion in time.

The Roots of Energetic Abstraction

Bogdanov had already noticed in passing the similarity of the form of the energy concept and abstract labour. Both concepts abstract from the concrete activities of human beings or natural forces; they grasp them only from a quantitative point of view – from the amount of work they can perform per unit of time. This is why, in Bogdanov's view, they could substitute monetary valuation: When calculation in energy or abstract labour replaced calculation in value, the rule of exchange over use value would be broken. Labour, a transformative capacity, was the most basic use value in Bogdanov's understanding of socialism. For Bogdanov, it was still possible to identify the abstraction of nature's work and human labour, historically brought about by capitalism, with a 'true' experience of nature. While the 'labour point of view' was not a transhistorical truth to Bogdanov, it was historically true as an experience that could guide socialists in their realization of a proletarian republic of labour.

The Soviet economist Isaak Rubin, now considered a pioneer of value theory, criticized Bogdanov for identifying energy with value. The physical forms of production do not coincide with the social forms of capitalism and cannot be substituted for them. Emphasizing that Marx understood value as a social phenomenon, he wrote:

> One of two things is possible: if abstract labor is an expenditure of human energy in physiological form, then value also has a reified-material character. Or value is a social phenomenon, and then abstract labor must also be understood as a social phenomenon connected with a determined social form of production. It is not possible to reconcile a physiological concept of abstract labor with the historical character of the value which it creates.[30]

That the 'physiological expenditure of energy as such is the same for all epochs' and that one might then say that 'this energy created value in all epochs' makes the energy concept useless for a historically specific understanding and critique of capitalism. It is questionable whether this really captured Bogdanov's understanding of energy, which, as we have seen, had a historical dimension, insofar as the 'labour point of view' emerged only under machine production. To Rubin, in any case, energy – or any physiological equality of all labour – was a *presupposition* of abstract labour, not its content.[31]

The concept of abstract labour marks the nexus between value and commodity production. Commodities have a double character as both use values and exchange values. In the value form, they 'acquire the new determination of exchange value which abstractly negates all differences of use value between commodities and thereby declares them all identical as values'.[32] Through this identity in terms of value, the labour that produced those commodities, whether living or congealed in

30 Isaak I. Rubin, *Essays on Marx's Theory of Value* (Black Rose Books, 1973), 135.

31 Ibid., 137–9.

32 Christopher J. Arthur, 'Value, Labour and Negativity', *Capital and Class* 25, no. 1 (March 2001): 19.

technologies, becomes related as well and reduced to abstractions of themselves – at least for capital. This is so because capital has an interest in exploiting all labours – and employing all use-value-creating nature one might add – regardless of their concrete specificity.[33] Abstract labour is labour seen from capital's point of view, and it entails the abstraction of nature in production.

Since Rubin's work has been rediscovered and translated in the 1970s, this debate on the nature of abstract labour, and the social forms of capitalism more broadly, has been revived – and energy still plays a role in it. Some have argued that for labour and nature to be abstractable in this way requires them to share a material quality. This physiological quality – the expenditure of 'energy' or 'human corporeality' – constitutes the quantitative, abstract side of all human labouring and its metabolic relationship to other parts of nature.[34] Only because of this shared material nature can animals, human beings, water streams, and burning coal substitute and act on each other. In this understanding, energy takes the role of the substance of transhistorical abstract labour. Others hold that 'abstract labour' has no foundation in nature but is the social form that labour takes under capitalism, an appearance of pure motion in abstract time and space.[35] It would be a mere 'capitalist prejudice' to treat 'living labour as an energetic economic resource'.[36] Despite this debate, there have been no attempts by either side to understand *how* the energetic abstraction *has* historically emerged within capitalism. And

33 Ibid., 19–21.

34 Guido Starosta and Axel Kicillof, 'On Materiality and Social Form: A Political Critique of Rubin's Value-Form Theory', *Historical Materialism* 15, no. 3 (1 January 2007): 20. It should be noted that Starosta and Kicillof talk no longer of 'energy' but of 'corporeality' in their most recent texts. This might suggest they have moved away from assuming a given, measurable physiological basis, but want to emphasize a material continuity between the subject of labour and its object. Why this continuity would be called abstract (rather than 'abstractable') is another question.

35 Werner Bonefeld, 'Abstract Labour: Against Its Nature and on Its Time', *Capital and Class* 34, no. 2 (1 June 2010): 257–76; Werner Bonefeld, 'Debating Abstract Labour', *Capital and Class* 35, no. 3 (1 October 2011): 475–9; Arthur, 'Value, Labour and Negativity'.

36 Bonefeld, 'Debating Abstract Labour', 478.

where the concept of abstraction is invoked in the interdisciplinary field of energy history, it is rarely linked to debates in value theory.[37]

This raises the question of the origin of abstract forms of knowledge within capitalist society. On a general level, approaches differ in whether they deduce the categories of thought from the sphere of production or circulation. While the former focuses on technologies and the division of labour, the latter emphasizes commodity circulation and abstract equivalence. Bogdanov himself can be seen as a pioneer of the former and of the 'external' explanation offered by Soviet scientists and historians – remembered as the Hessen-Grossmann thesis – which left a definite mark on British historians of science.[38] For him, the concept of energy emerged in machine production. The Dutch council communist Anton Pannekoek, himself an astrophysicist, defended Bogdanov against Lenin's correspondence theory of truth: Energy was not 'discovered' in nature; it was formed from the observation of phenomena through the active thought-work of scientists.[39] Pannekoek did not, however, venture a hypothesis about the origin of this abstraction – about how that scientific work might be related to social praxis. Around the same time, the Institut für Sozialforschung in Frankfurt commanded a study by Franz Borkenau on the social origins of the mechanical worldview, which was considered a failure as it did not take the technical-scientific development sufficiently into account.[40] Inspired by György Lukács, others derived the

37 On Barak, 'Three Watersheds in the History of Energy', *Comparative Studies of South Asia, Africa and the Middle East* 34, no. 3 (1 January 2014): 440–53; Philip Mirowski, *More Heat Than Light: Economics as Social Physics, Physics as Nature's Economics* (Cambridge University Press, 1989).

38 Gerardo Ienna and Giulia Rispoli, 'The 1931 London Congress: The Rise of British Marxism and the Interdependencies of Society, Nature and Technology', *HoST: Journal of History of Science and Technology* 15, no. 1 (1 June 2021): 107–30.

39 Anton Pannekoek, 'Lenin as Philosopher: A Critical Examination of the Philosophical Basis of Leninism', *Philosophy and Phenomenological Research* 10, no. 1 (September 1949): 144.

40 Franz Borkenau, *Der Übergang vom feudalen zum bürgerlichen Weltbild: Studien zur Geschichte der Philosophie der Manufakturperiode* (Wissenschaftliche Buchgesellschaft, 1980). Borkenau's study could not fulfil the institute's expectations; see Henryk Grossmann, 'The Social Foundations of Mechanistic Philosophy and Manufacture', *Science in Context* 1, no. 1 (March 1987): 129–80. Nevertheless, an

forms of knowledge more from the sphere of circulation and the equivalence of commodities on the market. Alfred Sohn-Rethel, a student of Ernst Cassirer, whose work on the concept of energy we have encountered above, dedicated his life to the project of deriving the Kantian transcendental subject – the *a priori* of time, space, and causality – from the real abstraction performed in commodity exchange. Sohn-Rethel tried to show that the faculty of abstract thinking, 'of seizing what is common to several objects without being visible in any of them', was rooted in commodity exchange.[41] While some scholars speculated that this could be made fruitful for an analysis of the energy concept, Sohn-Rethel himself never made this connection explicitly.[42]

Under fully developed capitalism, however, there is no neat distinction between a sphere of circulation and a sphere of production – they mediate one another. All production is *commodity* production; and commodity production not for 'equal' exchange but for profit.[43] Not only does exchange on the market transform useful products into 'bearers of social alienation', but the productive activity itself is already alienated, formed according to the needs of accumulating capital.[44]

---

interest in the interrelation between economy and intellectual forms remained present in the institute's work. However, it was posed less as a question of historical-materialist analysis and more as a question of what a Marxist epistemology between Kant, Hegel, and Marx might look like – a project Alfred Schmidt shared with French Marxists like Louis Althusser. Alfred Schmidt, ed., *Beiträge zur marxistischen Erkenntnistheorie* (Suhrkamp, 1969); Alfred Schmidt, *The Concept of Nature in Marx* (Verso, 2014).

41 Anselm Jappe, 'Sohn-Rethel and the Origin of "Real Abstraction": A Critique of Production or a Critique of Circulation?', *Historical Materialism* 21, no. 1 (1 January 2013): 3–14.

42 Carl Freytag, 'Alfred Sohn-Rethel, die Vorsokratiker und die kritische Liquidierung des Apriorismus', *Recherches germaniques*, no. HS 15 (10 July 2020): 70–9; Mirowski, *More Heat Than Light*, 139; Herbert Breger, *Die Natur als arbeitende Maschine: Zur Entstehung des Energiebegriffs in der Physik, 1840–1850* (Campus Verlag, 1982), 19.

43 Frank Engster and Oliver Schlaudt, 'Alfred Sohn-Rethel: Real Abstraction and the Unity of Commodity-Form and Thought Form', in Beverley Best, Werner Bonefeld, and Chris O'Kane, eds, *The SAGE Handbook of Frankfurt School Critical Theory* (SAGE Publications, 2018), 295.

44 Jappe, 'Sohn-Rethel and the Origin of "Real Abstraction"', 8.

Commodity production by large-scale machinery entails the energetic measurement of both human and natural work, to subsume it under a common value, compare it, and replace it. 'In a world dominated by exchange value', writes Amy Wendling, 'humans and machines are interchangeable, and both are subject to an abstract calculation of their quantifiable value.'[45] Thus, the epistemological question cannot be isolated from any of the two spheres. Technical development cannot be treated as independent, but neither should it be reduced to the implementation of abstract forms of thought resulting from exchange.[46]

The Objectification of Energy

What historical conditions allow for the treatment of nature as a working entity? The concept of energy emerges from social practice and machine production for the market: It refers to the problem of putting natural forces to work and shaping them into regular, flexible sources of power. A way to think about the appropriation of natural forces as sources of power is in terms of *formal or real subsumption of production*. This distinction marks the depth to which capital transforms the labour process. Formal subsumption 'does not in itself imply a fundamental modification in the real nature of the labour process', whereas real subsumption denotes the material and social transformation of production into forms 'more adequate to capital for the simple reason that they

45 Wendling, *Karl Marx on Technology and Alienation*, 11. François Vatin and Amy Wendling even see this energetic understanding of human and natural labour as the basis of Marx's concept of surplus value and, hence, as the condition of possibility to diagnose exploitation. The argument is not that energy is the substance of value, as Podolinsky argued, but that the diagnosis of exploitation presupposes an abstraction from qualitatively different labours analogous to the concept of energy; Vatin, *Le travail, économie et physique*, 105–22; Wendling, *Karl Marx on Technology and Alienation*, 81.

46 Katherina Kinzel and Jan Overwijk, 'Marxism Without Irony: On the Failure of Alfred Sohn-Rethel's Transcendental Materialism', *Constellations*, 21 September 2025, 6.

press out more surplus-value'.[47] Capital thus creates the conditions for continued capitalist production. The development of machinery – an apparatus consisting of motor mechanism, transmission mechanism, and working tool – is a manifestation of the real subsumption of labour. Some doubt that there was ever a historical period of only formal subsumption, where production for capital accumulation has not somehow affected the labour process. Still, the distinction is analytically useful to trace qualitative changes in production such as the emergence of an 'energy part' of the means of production.

The energy economy has its origins in the *mechanization of the motor* which is a moment in the development of the machine system. The motor becomes mechanized and objective, when it is possible to distinguish a 'mover' from the 'apparatus it moves', when 'movers' are substituted for each other and fitted to the productive apparatus with the purpose of producing more competitively. In Marx's account, the mechanization of the tool precedes and triggers the mechanization of the motor and the employment of natural forces.[48]

> It was . . . the invention of machines that made a revolution in the form of steam engines necessary. As soon as man, instead of working on the object of labour with a tool, becomes merely the motive power of a machine, it is purely accidental that the motive power happened to be clothed in the forms of human muscles; wind, water, or steam could just as well take man's place.[49]

Just as well, or even better: Living beings are lousy performers of uniform, regular motive power. The motor mechanism becomes reshaped and

47 Marx, *Capital*, vol. 1, 1021; Patrick Murray, 'The Social and Material Transformation of Production by Capital: Formal and Real Subsumption in *Capital*, Volume 1', in R. Bellofiore and N. Taylor, eds, *The Constitution of Capital: Essays on Volume 1 of Marx's 'Capital'* (Springer, 2004), 257.

48 Later historical work confirms Marx's broad-stroked outline of the development of large-scale machinery and the uptake of the steam engine in British industrialization; Nick von Tunzelmann, *Steam Power and British Industrialization to 1860* (Clarendon, 1978), 8; Malm, 'Marx on Steam', 176.

49 Marx, *Capital*, vol. 1, 497.

remade in a way that its capacity, regularity, and flexibility to perform work are increased and its cost of power decreased.

It is difficult, if not impossible, to pin this to a historical moment or resource. Andreas Malm suggests that the shift from water power to steam is the shift from the mere formal subsumption of the flow to the real subsumption of the coal stock. 'The subsumption of this force of nature' – the flow of water – 'is bound to be ever formal . . . Only the stock [of coal] can be conjured up as a power in motion internal to capital itself.'[50] Yet it is difficult to insist on an absolute distinction between water and coal-fired steam power. Water power was the first power source for large-scale industry. Where it was readily available, it could remain an alternative to steam, in some places well into the twentieth century. As early as the eighteenth century, water-powered infrastructure already powered mines and mills, much of which involved significant alterations to rivers, the digging of reservoirs to overcome seasonal variations, and, more generally, a 'rationalization' of the water economy. The waterwheel was also the experimental system in which engineers first coined the mechanical concept of work (*travail*) as the effect of a force acting through space – a concept which in turn influenced the design of waterwheels.[51] In the theory of the steam engine, Sadi Carnot drew on the mechanical concepts that had been developed to understand water powered apparatuses. The transition from water to steam is less a rupture than a gradual process of shaping the motor mechanism for the purposes of large-scale industry.

Yet one cannot ignore the marked difference in subsumption indicated by the shift to steam as the *machine-production of working nature*. Steam is the first material that is commanded into existence to power machinery.

50 Malm, *Fossil Capital*, 312.

51 François Vatin, 'Le "travail physique" comme valeur mécanique (XVIIIe–XIXe siècles)', *Cahiers d'histoire: Revue d'histoire critique*, no. 110 (1 October 2009): 117–35; Andrew M. A. Morris, 'John Smeaton and the Vis Viva Controversy: Measuring Waterwheel Efficiency and the Influence of Industry on Practical Mechanics in Britain 1759–1808', *History of Science* 56, no. 2 (June 2018): 196–223; K. Chatzis, 'Économie, machines et mécanique rationnelle: La naissance du concept de travail chez les ingénieurs-savants français, entre 1819 et 1829', *Annales de ponts et chaussées* 82 (1997): 10–20; Terry S. Reynolds, *Stronger Than a Hundred Men: A History of the Vertical Water Wheel* (JHU Press, 1983), ch. 5.

Rather than transmitting a movement, as wheels, belts, and pulleys had previously, the steam engine creates an 'artificial' movement in a closed vessel. By making use of its elastic properties – steam's capacity to expand when heated and condense when cooled – the engine harnesses a movement it has itself produced. By mediating fuel's combustion and the apparatus onto which the piston's movement is transmitted, steam allows for a flexibility of inputs and outputs. Because all biomass, combusted under a boiler, causes water to evaporate, steam engines can run on a variety of fuels. They can also be operated at a variety of speeds and regulated by pressure and temperature. Steam made the location, volume, and temporality of power more controllable and reproducible – in Sadi Carnot's words, steam was a 'universal motor' and potentially a general condition of production.

The mechanical production of working nature introduced the possibility of a concentration and capitalization of mechanical power. This first took the form of steam engine firms and the manufacture of patented components of engines. The steam economy meant widespread circulation of fuels, engines, and machine tools such as the power loom; steam and mechanical power themselves were not generally traded as commodities.[52] Steam was a working fluid whose volume and temperature stood in a quantitative relationship to the output of production. In its mediation of fuel's combustion and the machine's movement, ownership was not altered. It remained legally and spatially immobile. This meant that consumers of power could purchase or rent the means of steam production so as to generate motive power in situ at any time, but they could not rent or buy power as such.

Electricity altered the above conditions and thus qualifies as the second and indeed more profound form of working nature. The electric

52 Early on in the British steam economy, it was not uncommon for power to be rented to small manufacturers. S. R. H. Jones describes the cottage industry in Coventry, where individual producers bought power from a central steam engine, priced according to the number of power looms they employed. See S. R. H. Jones, 'Technology, Transaction Costs, and the Transition to Factory Production in the British Silk Industry, 1700–1870', *Journal of Economic History* 47, no. 1 (March 1987): 90.

current is a technical fact that nature itself does not produce; it exists only in a circuit or grid. The emerging electrical power industry of the late nineteenth and early twentieth centuries was more concentrated than the steam engine industry. Linked by patents, a small number of firms were capable of generating electricity through turbines, generators, transformers, and other grid technology and thereby controlled the construction of power stations globally.[53] Because electricity can be transported only across a grid, its market and infrastructure are virtually identical. Regulated as a 'natural monopoly', the operation of a power plant means access to all present and future consumption of a certain territory. What is more, working nature itself takes on a different form with electricity. The electric current is not limited to motive power but offers a more flexible form of nature's work. It illuminates, catalyses, coordinates, and transmits. Industrial and commercial consumers could now buy a precise amount of work: Electricity is standardized according to its voltage and frequency and divided into units of what constitutes its use value – the kilowatt-hour (kWh), or work performed over a period of time. In the power economy, natural forces are appropriated and transformed into a product that comes closest to an energy commodity.

Machinery's advantage was that power supply could be combined with the rhythm, volume, and location of capitalist production. With the abandonment of water power and the development of steam and electricity, however, the uncertainty of nature's behaviour did not vanish but was displaced. The control of nature's work ends where coal is heaped into the furnace's mouth, as every delivery of coal differs in its calorific value. It ends in the mine, where every seam, every bulk of coal, exhibits slightly different features. Due to their different aggregate

53 Vaclav Smil, *Energy and Civilization: A History* (MIT Press, 2017), 386; William J. Hausman, Peter Hertner, and Mira Wilkins, *Global Electrification: Multinational Enterprise and International Finance in the History of Light and Power, 1878–2007* (Cambridge University Press, 2008), 79, 92–5. See, for the comparably local steam engine business, in which the pirating of engine design was common, Peter Temin, 'Steam and Waterpower in the Early Nineteenth Century', *Journal of Economic History* 26, no. 2 (1966): 189–90.

states, this problem is less pronounced with petroleum, gas, and hydroelectricity. Yet an uncontrolled variation in the quality of natural forces persists.

The production and control of working nature are impossible without intellectual labour. Through the work of engineers, scientists, and managers, energetic abstraction is brought to bear upon natural materials, and the productive apparatus is made to approximate pure motion through time and space. As Harry Braverman has argued, abstract labour exists as an idea 'in the mind of the capitalist, the manager, the industrial engineer'.[54] I call this third moment *energetic determination* or *energetic valuation*, the production of calculations, measurements, and statistics undertaken with the purpose of controlling or improving nature's performance of work. It yields the form nature's changeability takes in capitalist society.[55]

Such abstract forms of knowledge have typically been understood as preceding the rationalization of production. In the early history of the energy economy, however, the use of machines – and the treatment of different sources of power as equivalent – enables the scientific understanding of their equivalence. The use of waterwheels and steam engines predates energy science; in fact, it can be argued that the latter emerges primarily from the former.[56] Because fuels appear with no indication of

54 Harry Braverman, *Labor and Monopoly Capital: The Degradation of Work in the Twentieth Century* (Monthly Review Press, 1974), 22.

55 Energetic valuation should not be misunderstood for a tendency of capitalists to choose the most energetic material – that is, fuels of a higher calorific value. All calculation of the *cost of power* implies a *calculation of power*, regardless of the cost for which the capitalist in the end settles. A cheaper prime mover is only *one way* to bring production costs down. Malm's study is only one indication of the indirect effect on other costs due to changes in power technologies, by way of 'deskilling' in labour or the factory.

56 While the intellectual contribution to nature's work ultimately remains rooted in production it does not take place only in and around production. Even prior to industrialization, nature appears in light of it – in rivers that are not yet tapped, coal reserves that are not yet mined, or territories not yet surveyed. The French and Prussian states of the early nineteenth century supported scientific research so as to facilitate industrialization and 'catch up' with Great Britain. Understood as a concrete historical process, there may well be a 'theoretical'

their value as energy, they are, for a time, consumed without knowledge of 'how much' they work. This even holds true for the steam in the first engines running mills (rather than pumping water): Their power could in the beginning only be adapted to the productive apparatus through trial and error.[57] Theoretical and practical mechanics merge only in the nineteenth century, just to split again into formalized physics and the 'applied' sciences around the turn of the twentieth century: fuel science, mechanical and electrotechnical engineering, and so forth. As nature's capacity to work is not a thing to be detected in natural materials, but a relation between those materials and machine, a performance that unfolds within a technical apparatus, energetic determination never comes to an end.

The objectification of energy finds its limit in nature. However much the motor mechanism is subsumed, however fossilized or electrified the energy economy becomes, abstract labour can never be realized. Nature is not so thoroughly 'abstractable' that it can ever become abstract; there is always a residual negativity. Scholarship on fossil capitalism sometimes tends to reproduce the capitalist abstraction of fossil fuels. When ecological historians like Wrigley and Sieferle argue that the *energetic* advantage of coal together with an *energetic* need of society created the first instance of a fossil-fuelled economy, proponents of fossil capitalism point out that it was the *abstract nature* of fossil fuels together with a capitalist economy striving towards abstraction that brought about fossil capitalism. Elmar Altvater describes the congruence of fossil energy's

---

rationalization of production that precedes industrialization. Intellectual work that transcends the single factory, surveying, comparative analysis, and standardization become a necessary part of the forces of production. Ursula Klein, *Technoscience in History: Prussia, 1750–1850* (MIT Press, 2020); I. Grattan-Guinness, 'Work for the Workers: Advances in Engineering Mechanics and Instruction in France, 1800–1830', *Annals of Science* 41, no. 1 (January 1984): 1–33. This imperial competition is present in the works on political economy and even in the forewords of machine treatises like Sadi Carnot's. In 1827, Charles Dupin calculated and compared the French and British 'productive forces' to assess their relative power; Charles Dupin, *Forces productives et commerciales de la France* (Bachelier, 1827).

57 Richard L. Hills, *Power from Steam: A History of the Stationary Steam Engine* (Cambridge University Press, 1993), 88–9.

properties with the 'logics of capitalist development'.[58] Along similar lines, Malm argues that in using coal, 'capital dug into a source *whose most concrete quality was abstractness*'.[59] To describe something physically existing as 'abstract' suggests that nature is prepared for capitalist abstractions, that fossil fuels are 'proto-capitalist' by nature. While one might say that nature allows for an (always relative, temporarily limited) approximation of pure motion in time and space, this does not make it abstract.

Fossil fuels comprise a wide range of properties. Their natural history did not prepare them for sale on the market or input into an abstracted labour process. As Alfred Sohn-Rethel put it, commodification is a claim *against* nature 'to abstain from any ravages in the body of this commodity and to hold her breath, as it were, for the sake of this social business of man'.[60] Yet it never fully succeeds in doing so. At any moment in the history of production, certain qualities make objects more or less amenable to capital accumulation. Yet despite historical change, the distinction between the abstraction capital seeks to realize and nature's own particular forms never collapses. Nature, including fossil fuels, is 'always "in and against" capital'.[61]

Residual Negativity

The difference between forms of nature and abstract labour – the way capital 'cognizes' nature – is categorical and does not concern the question of real subsumption. Although capitalist production approximates capital's perception of production (that is, as pure movement or change in time and space) this does not mean that human labour and working nature ever take the form of pure change in time. There is a remaining source of tension, a residual non-identity, between the way in which

58 Altvater, 'The Social and Natural Environment of Fossil Capitalism', 41.

59 Malm, *Fossil Capital*, 302. Emphasis mine.

60 Alfred Sohn-Rethel, *Intellectual and Manual Labour: A Critique of Epistemology* (Brill, 2020), 21.

61 Arthur, 'Value, Labour and Negativity', 29.

capital organizes production and the things it organizes. Echoing Adorno's negative dialectics, Christopher Arthur argues that there is a categorial distinction between the physical forms of production (the object of an analysis of real subsumption) and the abstraction of labour (the object of value theory). There may be a simplification of the labour process along the lines of the social form, but this should 'be understood as merely an approximation to the "content" of the concept of "abstract labor"'. If the labour process is streamlined and homogenized, 'even the simplest motion still has some quality, it can never be abstraction as such'. This is why, in studying real subsumption, 'the distinction between abstract and concrete cannot be collapsed'.[62] Nature cannot be positively identified with value.

The capitalist cannot organize the objective world – labour, resources, and machinery – as she wishes. Its organization as forces of production reflects capital, comes to resemble it and mimics its abstract motion, but concreteness is also a limit. The form of capital and its content retain 'a structure of abstract contraposition: the content is inscribed in the form while retaining much that cannot be grasped in it'.[63] This remaining negativity is not just a feature of that which is not produced, as Karl Polanyi once argued, nor is it limited to living beings or conscious subjects. All natural things stand in relations that go beyond what the value form captures and employs. A moment of resistance can be found even in the most sophisticated machineries; the most naturalized concepts, such as energy, generate friction when they are made to bear on objects. As we will see later in the book, it is precisely the purest form of energy, and the one in which most human labour has sedimented – the electric current – which exhibits such a peculiar materiality.

While human labour is more flexible and malleable than other parts of nature, it is also more prone to resistance. Labour is a faculty tied to human life and its subjectivity, a conscious relation to oneself which is socially and culturally formed. This is why labour has a frustrating autonomy and obstinacy (*Eigensinn*), a capacity to free itself from even

62 Ibid., 21 2.

63 Ibid., 29.

the most fully subsumed production line.[64] Its capacity for self-regulation, a residual subjectivity of labour, is a constant threat to capitalist production. Human beings always find reasons to waste time and avoid work; cooperation can be withheld. Under conditions of alienated labour, the worker's subjectivity shows itself less in the product she produces than in its resistance – spontaneous or planned, single or coordinated – against the labour process imposed on her. Without labour's negativity, the capitalist imperative to discipline labour, and the search for ever more effective means to do so, could not be understood. When buying labour power, there 'is no guarantee that the buyer actually gets what he pays for'.[65]

Labour differs from other factors entering production, but not in an absolute way. To highlight the specificity of human labour, nature's own negativity has often been underplayed. Arthur argues that the worker's subjectivity poses a particular problem for capital 'because it gives rise to a definite recalcitrance to being "exploited" which the other factors do not possess'.[66] Resources, land, and materials cannot let their thoughts wander, they do not carry over emotions from a life outside the factory, and they cannot make their exploitation a source of resistance. Braverman suggests that labour and labour power are barely distinguishable in animals and non-animated parts of nature, as 'the most cunning contrivances can get from the labour of the animal only minor variations of actual labour'.[67] Along these lines, Arthur claims that resources and machines enter the production process 'with their productive potential given, known in advance'.[68] Given the long history of breeding and genetic modification, and of the role of production in fostering knowledge of materials, this seems at best empirically reductive. In any single production process, there may be a tendency for bred animals, constructed machinery, and processed material to behave as the breeder,

64 Malm, *Fossil Capital*, 309; Alexander Kluge and Oskar Negt, *History and Obstinacy*, illustrated ed. (MIT Press, 2014).

65 Malm, *Fossil Capital*, 309.

66 Arthur, 'Value, Labour and Negativity', 30.

67 Braverman, *Labor and Monopoly Capital*, 55.

68 Arthur, 'Value, Labour and Negativity', 30.

engineer, and chemist intended – but as a condition of production, the capitalist cannot control them.

A more fundamental problem with this argument is that it conceives of nature's work and human labour undialectically, as entirely distinct activities – when they are in fact mediated through one another. Alexander Kluge and Oskar Negt speak of a capacity for self-regulation encompassing nature and human beings that cannot be fully suppressed. Nature's resistance to abstract valuation cannot be entirely bracketed; the abstraction can never be fully realized, the object never entirely identified with the subject. Nature cannot be shaped at will; there are properties of nature, including in the nature of human beings, that resist subsumption. Their peculiarity is precisely in how they contradict attempts to socialize them. Yet this counter pole of capital exists only as negativity, is united only in its natural and human resistance; it cannot be affirmed and has no logic or economics of its own.[69]

Working nature and human labour are also historically related. The more durable forms into which scientists and engineers can mould inanimate nature become a means to intensify the exploitation of labour. This holds true for the objectification of the prime mover, the streamlining of natural forces into sources of energy, and the emergence of the energy economy more broadly. The outsourcing of physical labour leads, first, to a deskilling and employment of cheaper labour; second, to an exploitation of differences in labour, through temporal and spatial independence; and, third, to an intensification of exploitation through an

69 Kluge and Negt, *History and Obstinacy*, 85, 113. Still, to sketch such an economics of human labour capacity is exactly what Kluge and Negt set as their task: They search for self-regulation that 'manages to incorporate the overabundance of primary regulations, which elude human will in itself and remain alien to it, into the relationality of living labour without resorting to any exclusionary mechanisms. When this succeeds to a sufficient degree, it gives rise to an autonomous force field that occurs naturally neither in nature nor in history as an enduring condition. This is the meaning behind the phrase the "naturalism of man and the . . . humanism of nature".' Simon Schaupp has recently begun to explore how workers use nature's resistance in their labour struggles, what he calls 'ecological obstinacy'; Simon Schaupp, *Stoffwechselpolitik: Arbeit, Natur und die Zukunft des Planeten* (Suhrkamp, 2024), 49–54.

increased control of the pace of production. Even the greater dependence on human labour in the production of energy was only temporal and was reduced by way of labour-saving forms of energy, such as oil and electricity, along with general automation.[70] This continued with electrification and the use of renewable energy.[71]

The following chapters inquire into the effects the abstraction of labour generates when it shapes the stuff of nature into a 'working nature'. What is negated and what is affirmed when natural forces are shaped into sources of energy? In their objectification as sources, the transformative capacity of natural forces is not neutralized but appropriated in an altered form. Natural forces are embedded in technology as they are shaped and constrained by technology. Energy transforms the world using inherent forces of nature but in a truncated, restricted form. In the energy concept, nature's changeability appears as controllable, mobile, calculable units for sale. Energy is thus constituted through the dialectical overcoming of natural forces, which are both negated (in their excessiveness, abundance, irregularity, and multi-directionality) and preserved as the 'substance' of energy resources, their objective ability to perform work.

On the surface, this book can be read as a history of abstraction, from individual resources like coal and infrastructures like the power grid to the 'representation' of the world energy economy in a single energy balance. While there are historical connections between the chapters – the electricity industry pushes for a more precise definition of calorific value, and data on coal and electricity go into energy balances – these moments of abstraction do not add up to a single system. While the unit of energy in global energy balances may be the most abstract (it aims to subsume all commercial energy sources worldwide), it remains just that: a practical abstraction for the purposes of governance and investment. The moments of abstraction do not form a succession towards an

70 Malm, *Fossil Capital*, 313; Mitchell, *Carbon Democracy*.

71 Rebecca Pearse and Gareth Bryant, 'Labour in Transition: A Value-Theoretical Approach to Renewable Energy Labour', *Environment and Planning E: Nature and Space* 5, no. 4 (2022): 1878–80.

abstraction of energy that contains and sublates all previous natural and social resistances against it.

In other words, the movement presented here is not a positive but a negative dialectic: not a gradual sublation of natural resistance into higher forms of abstraction, but a chronicling of the residual non-identity, the parts of first and second nature that don't behave like energies, regular work per time, in production. This resistance is to be found not only on the fringes of the energy economy – let us say, in a study of peat, horses, and wood pellets – but at its very centre. As we will see in the next chapter, coal, this cherished 'stock of energy', was by no means a material whose productive power was known in advance and guaranteed in production. With the steam engine, the problem of controlling labour and nature's work in production reappeared on a different level.

3
The Value of Coal

There is no coal molecule.

– Franz-Josef Brüggemeier, Michael Farrenkopf, and Heinrich Theodor Grütter, *Das Zeitalter der Kohle*

Amid calls for a global coal phase-out, an international coal market has emerged largely unnoticed. After two centuries of seaborne coal trade, the steam coal market, long divided into Atlantic and Pacific parts following different rhythms and prices, has now become largely integrated.[1] This means that the price of coal of a certain quality that is traded via long-term contracts or on the spot market is now roughly the same worldwide. Because coal is a complex and diverse material, and not every coal is suited

1 Rudianto Ekawan and Michel Duchêne, 'The Evolution of Hard Coal Trade in the Atlantic Market', *Energy Policy* 34, no. 13 (September 2006): 1487–98; Rudianto Ekawan, Michel Duchêne, and Damien Goetz, 'The Evolution of Hard Coal Trade in the Pacific Market', *Energy Policy* 34, no. 14 (September 2006): 1853–66; Raymond Li, Roselyne Joyeux, and Ronald D. Ripple, 'International Steam Coal Market Integration', *Energy Journal* 31, no. 3 (2010): 181–202; Aleksandar Zaklan et al., 'The Globalization of Steam Coal Markets and the Role of Logistics: An Empirical Analysis', *Energy Economics* 34, no. 1 (January 2012): 105–16; A. Wegerich, 'Digging Deeper: Global Coal Prices and Industrial Growth, 1840–1960' (DPhil thesis, University of Oxford, 2016).

for all purposes, international trade rests on an agreed-upon classification and standard of quality. On the most general level, coals are sold either as thermal (or steam) coals to be burned in power plants or as metallurgical (or coking) coals that are used in steelmaking. Below this, coals are sorted into ranks of gross calorific value (GCV), the single most important marker of quality. The calorific value is a quantity that denotes the energy content of fuels and largely determines the temperature and amount of power that can be yielded from them. At the end of the coal economy, it seems the solid fossil fuel is now closer to being a global, homogeneous commodity than ever before.

The institutionalization of an international standard for coal has occurred much later and remained more limited than one would expect given the long history and sheer volume of coal production. The European Coal and Steel Community and the American Society for Testing Materials had formulated standards for the European and North American markets respectively in the mid-twentieth century. Nowadays, seven different standardization systems are commonly used, one of them being the standard of the International Standardization Organization (ISO) of gross calorific value (since 1976) and coal classification (since 2005).[2] Today, coal contracts specify guaranteed qualities according to such international standards which also stipulate the units, apparatus, and protocol of coal analysis. Sellers and buyers rely on testing companies, which offer to perform a great range of analyses according to different industry standards. Such standards also serve as an anchor for the trading of coal futures.

While the largest part of the coal mined is still consumed domestically and not directly subject to international contracts, standardization had effects beyond the coal it refers to. The size of the seaborne coal market is now at 1.3 billion tons per annum, meaning that a mere sixth of global coal production is covered by international standards.[3]

2 Paul Baruya, *Losses in the Coal Supply Chain* (IEA Clean Coal Center, 2012), 33.

3 International Energy Agency, *Coal 2023: Analysis and Forecast to 2026* (IEA, 2023).

However, standards can change industries in more subtle ways: A decade ago, the world's largest coal producer, Coal India, adopted the International Standardization Organization's GCV standard as benchmark for its domestic coal. Even though India does not yet export large amounts of coal, adopting the standard has had internal consequences: Domestic prices have begun to follow the world market and companies have an incentive to process and upgrade India's traditionally low-quality coals into fuels of a higher value.[4] The calorific value is more than a descriptive measure: Rather than representing a fixed quantity discovered in a lump of coal, it outlines a horizon of fuel development.

Environmental and ecological histories often describe coal as a 'stock of energy'.[5] In this understanding, the heating value simply *represents* an accumulated capacity to perform work. This chapter complicates this view of a ready-made stock of energy by focusing on how coal became energetically evaluated and how it resisted this valuation. Talking of a stock implies a spatial homogeneity and temporal stability of coal's potential to perform work, which does not exist by nature but is created socially. It remains precarious. *In space*, coal is highly heterogeneous at all scales, from the microscopic level of a piece of coal to the coal seams of the earth. Coal is made up of the remains of plant material that has been sedimented, deprived of oxygen by flooding and the layering of other material, and subjected to a sequence of biological, chemical, and geological processes. Geologists now distinguish between a diagenetic phase, in which the plant material loses water and is biochemically transformed into peat and lignite, and a geochemical phase, which begins when the material sinks further down into Earth's layers. Under

4 Sumantra Bhattacharya and Rachit Tiwari, 'Non-Coking Coal Pricing in India', *Economic and Political Weekly* 49, no. 3 (2014): 20–2; Rachit Tiwari, S. Bhattacharya, and Piyush Raghav, 'A Discussion on Non-Coking Coal Pricing Systems Adopted in Different Countries', *Vikalpa: The Journal for Decision Makers* 40, no. 1 (March 2015): 62–73.

5 E. A. Wrigley, 'Energy and the English Industrial Revolution', *Philosophical Transactions of the Royal Society A: Mathematical, Physical and Engineering Sciences* 371, no. 1986 (13 March 2013): 4; Rolf Peter Sieferle, 'The Energy System: A Basic Concept in Environmental History', in Christian Pfister and Peter Brimblecombe, eds, *The Silent Countdown* (Springer, 1990), 17.

increasing pressure and heat, chemical bonds break down and the internal structure of the material changes, leaving a heterogeneous, solid material that varies even within seams and bulks. On an abstract level, this process can be described as carbonization – the relative share of carbon increases in the plant material as other, volatile matter escapes. However, this process started at different times, proceeds at different paces, can be interrupted, and is never total – between 10 and 50 per cent of coal mass is made up of elements that are not carbon.[6] For this reason, coal could be called hyper-local: The natural history of a particular place has been inscribed in its chemical composition and structure. There is no one coal molecule, no one structure of bound carbon, oxygen, hydrogen, nitrogen, and so on, that all coals share.[7]

In time, coal's capacity to perform work declines and sometimes vanishes entirely. From the moment coal seams meet the atmosphere in the mine, and when coal is broken up and hauled from the ground, the conditions under which it forms and endures in nature are disrupted, and coal enters a creeping, incessant process of combustion (oxidation) and weathering. In a baffling phenomenon called 'spontaneous combustion', the use value of coals can collapse entirely, taking with them capital goods like ships, buildings, or machinery. While such coal fires and explosions happen only on occasion, oxidation always implies a change of the chemical condition of coal and a deterioration of its use value. Without proper stockpile management, some coals may lose up to 20 per cent of their heating value upon storage; depending on coal rank and climatic conditions, the loss is typically around 1 to 5 per cent from pithead to consumer.[8] To track the condition of commodified coal, a complex product of natural and social history, at all stages between mine,

6 Helge Wendt, *Kohlezeit: Eine Global- and Wissensgeschichte (1500–1900)* (Campus Verlag, 2022), 110.

7 Franz-Josef Brüggemeier, Michael Farrenkopf, and Heinrich Theodor Grütter, *Das Zeitalter der Kohle: Eine europäische Geschichte: Katalogbuch zur Ausstellung des Ruhr Museums und des Deutschen Bergbau-Museums auf der Kokerei Zollverein, 27. April–11. November 2018* (Klartext Verlag, 2018), 110.

8 Anne M. Carpenter, *Management of Coal Stockpiles* (IEA Coal Research, 1999), 26–7; Baruya, *Losses in the Coal Supply Chain*, 37–9.

market, and furnace is an impossible task. Coal never behaved as a docile 'stock'; its storage and transport posed problems that required close observation and management of its declining value.

Yet to speak of an energetic value in coal is more deeply flawed: Energy is not a substance in a thing. In the sciences, energy is the property of a system (physics) or reaction equation (chemistry). Even when this chemical reaction is intentionally set off in a furnace, combustion is always more than a function of the chemical properties of the fuel. It unfolds in a dynamic interaction with its environment; pressure, temperature, airflow, and the design of the furnace all condition coal's generation of heat. In the words of one of the leading British coal scientists of the early twentieth century, Clarence A. Seyler, the combustion of coal constituted 'a highly-complicated and destructive distillation, and is not capable of being represented by any simple chemical equation'. Reflecting on the disarray of coal analysis at the time, he concluded that the results of measuring combustibility 'cannot be expected to agree unless all use a standard method'.[9] To treat the calorific value as a property of the coal material means to assign what is the result of an interaction – the release of heat during combustion – to only one of its inputs. Aware of this conditioning of its own research object, coal scientists like Seyler searched for regularity and order in how coals behave *under standard conditions of combustion and analysis.*

Rather than assuming coal is stored energy, this chapter asks when coal came to be seen and treated as nature that performed work. As laid out in the preceding chapters, I conceive of energy as an emerging social relation between the natural material and the forces and relations of production that is objectified and projected back onto the natural material. The uptake of the steam engine is a crucial moment in this story: Here, for the first time, rather than borrowing a movement from nature, a material, or state of matter, is produced for the sole purpose of performing work. This in situ production of motion is what makes steam

9 Clarence A. Seyler, 'Classification of Coals and the Interpretation of Analysis', in Allan Greenwell and J. V. Elsden, eds, *Analyses of British Coals and Coke* (Chichester Press, 1907), vi.

power potentially universal and allows industrial production to abandon flowing waters. Yet steam's universality should not be hypostasized: It rests on the abundant availability of combustible biomass in the earth's crust. While steam engenders a substitutability of fuels from the beginning, I argue here that the energetic determination of coal has its roots in the productive pressure on steam, particularly in transport.

Coal Before Steam

Coal has a long human history before it powered, and helped forge, the first steam engines, locomotives, and ships. Coal seams reach the surface at many places around the world, albeit in different forms and concentrations. For centuries before industrialization, coal was a local fuel, useful in the household, and appreciated in brewing, lime-burning, salt-making, and other trades. If the locally available coal was low in sulphur, it could replace charcoal in ironmaking. In Belgium, Great Britain, and some regions of China, there developed a substantial exploitation, processing, and trade of coal long before the eighteenth century.[10] After Chinese coal mining declined in the twelfth century, the British grew into the world's largest by the sixteenth and seventeenth centuries. John Hatcher estimated that half the British coal mined before 1700 was used in the home, and about a third in industry and manufacturing.[11] Yet seventeenth-century British householders, brewers, and salt-makers understood themselves to be consuming not 'energy' or nature's work, but rather 'coale', or 'earth fuel', a soft rock that made fire.[12] This fire did

10 Franz-Josef Brüggemeier, *Grubengold: Das Zeitalter der Kohle von 1750 bis heute* (C. H. Beck, 2018); Robert Hartwell, 'A Revolution in the Chinese Iron and Coal Industries During the Northern Sung, 960–1126 A.D.', *Journal of Asian Studies* 21, no. 2 (February 1962): 153–62; John Hatcher, *The History of the British Coal Industry*, vol. 1, *Before 1700: Towards the Age of Coal* (Clarendon Press, 1993); John U. Nef, *The Rise of the British Coal Industry* (G. Routledge, 1932); Kenneth Pomeranz, *The Great Divergence: China, Europe, and the Making of the Modern World Economy* (Princeton University Press, 2000).

11 Hatcher, *History of the British Coal Industry*, vol. 1, 409, 420.

12 Nef, *Rise of the British Coal Industry*, 248.

not perform anything comparable to what human beings did when they worked; it did not work an object like water or windmills; it was a source of fire, heat, and light.

Mines have long been seen as early sites of an abstract valuation of nature.[13] While British law stipulated that the right to mine coal lay with the owner of the surface, the continental *Bergregal* secured the state the rights to all subsurface resources – a privilege, however, that was not necessarily claimed with regard to coal.[14] Private investment for profit was common in agriculture and trade as well, but due to the high costs of underground mining, mines were among the first joint investments. In fact, the German word for union derives from the shareholder organization of a mine, the *Gewerkschaft*. Where landowners did not work their own mines, partnerships and joint stocks were common in Britain as well.[15] The borders between miners and owners or leaseholders were less clearly marked in the early industry. With rising costs of operation, absentee ownership and wage labour became more common.[16] Mining towns were among the first places where a vast majority of the population depended on the market for their livelihoods.[17] In Great Britain and elsewhere, coal mining oriented itself towards trade in coal as a commodity for the market.

The market for those coals was usually limited to some ten kilometres. Coal was a bulky good that did not fetch a high price per weight and crumbled during transportation. Before railways, water was the only

13 Lewis Mumford, *Technics and Civilization* (University of Chicago Press, 2010), 74–7; Jeannette Graulau, *The Underground Wealth of Nations* (Yale University Press, 2019).

14 Under conditions of fragmented political sovereignty, regulations varied widely. The exploitation of coal was not everywhere considered mining. Michael Walter Flinn, David Stoker, and Roy A. Church, *The History of the British Coal Industry: 1700–1830, the Industrial Revolution* (Oxford University Press, 1984), 36; Brüggemeier, *Grubengold*, 39.

15 Flinn, Stoker, and Church, *History of the British Coal Industry*, 39–41.

16 Brüggemeier, *Grubengold*, 33–8; Mumford, *Technics and Civilization*, 75.

17 Andreas Friedolin Lingg, *Die Entdeckung der Wirtschaft: Der mittelalterliche Bergbau und die Vermehrung der Welt* (Konstanz University Press, 2023), 199–208; Richard Dietrich, *Untersuchungen zum Frühkapitalismus im mitteldeutschen Erzbergbau und Metallhandel* (Georg Olms Verlag, 1991).

means by which coal could reach a distant market in quantity.[18] This is why London, supplied from the sea with Newcastle coal, could consume more coal than any other place by the eighteenth century, while wood dominated no more than fourteen miles from London. Contemporaries assumed that land carriage doubled coal's pithead price every ten miles, while coal could be carried by water twenty times as far for the same price.[19] Under these restricted transport conditions, the output of collieries was determined by the market they could serve.[20] Outside Great Britain, the Netherlands, and Belgium, collieries often had problems finding a market and rarely worked around the year before the mid-nineteenth century.[21]

Coal's uptake as a fuel has long been explained by a combination of its higher capacity to generate heat and a shortage of wood.[22] However, the relation is not so simple: Due to high transport costs, coal and wood remained restricted to local or regional markets, keeping shortages local as well. Complaints about a decline of forests were frequent, but historians have shed doubt on the existence of a *general* scarcity of wood in Great Britain or on the continent.[23] Invoking a looming catastrophe could serve other purposes, such as justifying the enclosure of forests and their use as an economic resource by the land-holder.[24] Coal's advantage as a fuel, self-evident from today's perspective, was not so self-evident either. Apart from price, what mattered in the early coal economy

18 Hatcher, *History of the British Coal Industry*, vol. 1, 13.

19 Flinn, Stoker, and Church, *History of the British Coal Industry*,146.

20 Hatcher, *History of the British Coal Industry*, vol. 1, 14.

21 Brüggemeier, *Grubengold*, 62–3, 68.

22 Nef, *Rise of the British Coal Industry*; Werner Sombart, *Die Genesis des Kapitalismus: Der moderne Kapitalismus*, vol. 1 (Duncker & Humblot, 1902); Rolf Peter Sieferle, *The Subterranean Forest: Energy Systems and the Industrial Revolution*, trans. Michael P. Osmann (White Horse Press, 2010).

23 Joachim Radkau, 'Zur angeblichen Energiekrise des 18. Jahrhunderts: Revisionistische Betrachtungen über die "Holznot"', *VSWG: Vierteljahrschrift für Sozial- und Wirtschaftsgeschichte* 73, no. 1 (1986): 1–37; Michael W. Flinn, 'Timber and the Advance of Technology: A Reconsideration', *Annals of Science* 15, no. 2 (June 1959): 109–20; Charles K. Hyde, *Technological Change and the British Iron Industry, 1700–1870* (Princeton University Press, 2019), 32–41.

24 Brüggemeier, *Grubengold*, 51.

was the quality of the flame, its burn, and its interaction with the manufactured product. Even though contemporaries were aware that coals could generate more heat than wood, they saw it as only one among many differences. Tim Nourse, who wrote an essay on London's coal use in 1700, estimated that 'one chaldron of coal would yield more heat than three or four loads of wood'. Despite this advantage, he preferred the cleaner-burning wood.[25] Because of its variance in burning, its smoke, and its smell, coal remained an unloved fuel that was avoided by households that could afford to.[26]

Coal was not generally preferred over wood in manufacturing either, but its higher yield of heat could be an advantage. In some fields, coal could easily substitute for wood, making its uptake a mere matter of availability and price. In smelting, the burning fire particularly mattered as it came in contact with the raw material. While a larger generation of heat could be beneficial in iron-smelting or glassmaking, both processes were sensitive to certain elements in coal, particularly sulphur, which were not present in wood and charcoal. Some problems of coal firing could be managed by redesigning the furnace and allowing for a better airflow.[27] Like in the household, heat development was not the only feature manufacturers looked for in a fuel. Anthracite, today praised for its high carbon content, proved practically inconvenient for most purposes in pre-industrial Britain, due to its lightness and resistance to catching fire. Once ignited, however, anthracite powder, or culm, produced a strong and equal heat appropriate for use in hothouses.[28]

This focus on fire, rather than heat, can also be seen from the methods to refine a coal's use value. Most methods were aimed at making the fire more agreeable, not necessarily hotter. Briquetting and coking were first introduced to satisfy the domestic consumer before they were applied in

25 Hatcher, *History of the British Coal Industry*, vol. 1, 38–40, 50.

26 Brüggemeier, *Grubengold*, 17–21.

27 Thomas S. Ashton, *Iron and Steel in the Industrial Revolution* (Manchester University Press, 1963), 37.

28 Nef, *Rise of the British Coal Industry*, 109–22; Hatcher, *History of the British Coal Industry*, vol. 1, 422, 425, 430, 441.

manufacturing.[29] Briquetting, which had long been common in Belgian colliery districts, was introduced to England at the onset of the seventeenth century. Sir Hugh Platt, who introduced briquetting to Great Britain, described a good coal as one that would melt during combustion and burn with a light fire. The making of briquets consisted of breaking up soft coals and mixing them with a pap of loam and water to form 'coal balls'.[30] Fire and fuel coincided to an extent that briquets were advertised as a 'new, delicate', and 'artificial fire'.[31] In the eighteenth century, the method of charcoal making was applied to coal: Coal was exposed to slow heating without burning it. During the process, the gaseous substances in coal evaporated and left a dense, baked product of coal, called coke. Coke yielded a pleasing fire whose reduced smoke did not corrupt metal. While smiths and ironworks had long used coal for some of the processes involved in making iron, they turned to coke only in the mid-eighteenth century.[32] Here, again, the advantage of coke was not primarily its larger heating value, but its reduction of substances that escaped during combustion and interacted with the metal.

When coal was traded between a mine and nearby manufacturers, or the city and its surroundings, as was the case before canals and rail were widespread, its use was confined both spatially and socially to those few lines exploring, mining, burning, and trading it. Except for long-distance trade limited to a few types of coal, sellers, traders, and buyers knew each other, and coals shaped their respective local furnaces. This symbiotic development of the early years is evident from the proliferation of names

29 Nef, *Rise of the British Coal Industry*, 246–51; Hatcher, *History of the British Coal Industry*, vol. 1, 414.

30 Hugh Platt, *A New, Cheape and Delicate Fire of Cole-Balles Wherein Seacole Is by the Mixture of Other Combustible Bodies, Both Sweetened and Multiplied. Also a Speedie Way for the Winning of Any Breach: With Some Other New and Serviceable Inventions Answerable to the Time* (Peter Short, 1603).

31 Nef, *Rise of the British Coal Industry*, 246–7; Barbara Freese, *Coal: A Human History* (Basic Books, 2004), 34.

32 Nef, *Rise of the British Coal Industry*, 251; Paul Mantoux and Thomas S. Ashton, *The Industrial Revolution in the Eighteenth Century: An Outline of the Beginnings of the Modern Factory System in England* (Methuen, 1964), 283–92, Ashton, *Iron and Steel in the Industrial Revolution*, 37.

for regional varieties of coal and the wide variation of measures between the coal regions.[33] Nef's history documented the customs of the British coal industry of the seventeenth century, in which collieries employed riddlers to sort mined coal into different types. Varying between the British coal valleys, the names for different types of coal expressed their use ('smith' or 'fire coal'), or reflected visible and palpable properties ('peacock' or 'cannel coal').[34] Coal was usually traded by a locally standardized weight or volume and not by a numerical scale of quality.

As long as coal was used in the household, manufacturing, and ironmaking, it remained conceptually and technically unconnected to mechanical power and human work. Domestic use brought about a knowledge of how coal burned and raised the issue of smoke and pollution, but it did not pose the question of coal's work. In manufacturing and ironmaking, a knowledge of how a coal burned and interacted with the manufactured product was crucial. Coals for ironmaking had to be rich in carbon and low in elements such as sulphur that impeded the strengthening of metal. Because coal entered ironmaking not only as heat energy but also as a raw material, coal's use value could not be easily quantified along a single dimension. Combustion alone does not necessitate a quantification of a coal's use value: The coal age engendered knowledge on fire, coal, and pollution, but not necessarily on energy – understood as the quantity of work performed in production. Before steam, coal's use value remained primarily qualitative: The aesthetics of fire trumped heat generation in most applications. To talk of pre-industrial 'energy' is thus always somewhat anachronistic.[35]

From the Element of Fire to the Substance of Heat

The equation of fire and fuel in the early use of coal reflected the chemical conception of combustion at that time: Fire was thought to be

33 Hatcher, *History of the British Coal Industry*, vol. 1, 46.

34 Nef, *Rise of the British Coal Industry*, 111.

35 Wendt, *Kohlezeit*, 109–44, 118; Brüggemeier, *Grubengold*, 43–53.

an element fixed within combustible substances and released during combustion. Eighteenth-century chemistry understood elements to confer properties onto substances and explain their different qualities and states. Those elements were initially thought to be detectable by their effect, not isolable in analysis. Over the eighteenth century, however, they came to be seen as 'obtainable in pure forms as products of analysis'.[36] The element of fire thus underwent a change from an elemental principle to a simple substance – the *calorifique* – that chemical analysis could determine. While the understanding of heat and combustion has changed since then, the term 'calorific value' still bears witness to this origin.

Theodore Porter argued that this shift was associated more with the art of mineral extraction and its economic interests than with alchemical theory.[37] While earlier mineralogist catalogues distinguished between metals according to features such as hardness, colour, and beauty, the new mineral assayers developed a taxonomy of 'simple substances'. Chemical investigations into those substances were directly linked to the exploitation of minerals. At that time, the Swedish crown and some local German authorities began to invest in mining academies, chemists, and mineral assayers, in an effort to increase their mining profits, a critical source of state revenue.[38] The knowledge produced in the Bergskollegium and Bergakademie was directly linked to the purpose of determining the amount of metal that could be commercially extracted from ore.[39] True to their task to improve the yields of minerals, these state mineralogists focused on metal and mineral ores like copper, gold, cobalt, and iron.

The isolation of 'fire' from wood or coal was not of the same interest to them, but fuels still found a place within the four classes of the

36 John G. McEvoy, 'Continuity and Discontinuity in the Chemical Revolution', *Osiris* 4, no. 1 (January 1988): 200.

37 Theodore M. Porter, 'The Promotion of Mining and the Advancement of Science: The Chemical Revolution of Mineralogy', *Annals of Science* 38, no. 5 (September 1981): 544, 547.

38 Ibid., 548–51.

39 Ibid., 547.

mineral kingdom: Inflammables (earlier called sulphurs according to an easily inflammable 'element') stood next to earths, salts, and metals, a classification reflecting the ancient distinction of water, earth, fire, and air.[40] Some chemists thought inflammables contained a firelike element, phlogiston, that accounted for their combustibility and was released in combustion. This element, however, could never be isolated in its pure form.[41] In the words of the Swedish mineralogist Torbern Bergman, it was 'so subtle indeed, that, were it not for its combination with other substances, it would be imperceptible to our senses'.[42] Bergman's description already reacted to the new imperative pioneered by practical mineralogists to identify elements and manipulate them.[43] Bridging a chemistry of principles and an emerging chemistry of simple substances, phlogistic chemistry never shared a unified body of theory but could take speculative or empirical forms, which sparked Antoine Lavoisier's statement that it was 'a veritable Proteus which changes its form every minute'.[44]

In the late eighteenth century, several chemists shed doubt on the theory that fire was a fixed matter in combustibles. As substances combined with fire took on a new form and sometimes increased in weight, Lavoisier reasoned, the fire could not reside within solid substances such as coal. Like Adair Crawford before him, he suspected that instead it was to be found in the gaseous states of vapour and air. In his view, inflammables were only the base of combustion: If the inflammable substance had a greater affinity to the part of the air that was not fire, and formed a compound with that part, the fire would be

40 Rachel Laudan, *From Mineralogy to Geology: The Foundations of a Science, 1650–1830* (University of Chicago Press, 1987), 24–6; Porter, 'The Promotion of Mining and the Advancement of Science', 552.

41 Ursula Klein and Wolfgang Lefèvre, *Materials in Eighteenth-Century Science: A Historical Ontology* (MIT Press, 2007), 150–1.

42 Torbern Bergman, *An Essay on the Usefulness of Chemistry, and Its Application to the Various Occasions of Life* (Murray, 1783), 102–3.

43 Robert Siegfried and Betty Jo Dobbs, 'Composition, a Neglected Aspect of the Chemical Revolution', *Annals of Science* 24, no. 4 (December 1968): 276; Porter, 'The Promotion of Mining and the Advancement of Science'.

44 McEvoy, 'Continuity and Discontinuity in the Chemical Revolution', 198.

set free.[45] The part of the air which readily combined with solid or fluid substances in combustion was identified as oxygen. Despite his avowed commitment to quantification by scales and balances, Lavoisier was no more able to weigh his *calorifique* than the proponents of phlogiston were to demonstrate the existence of fire matter.[46]

This chemistry of heat came with new instruments to measure and methods to calculate the heat produced in combustion. Lavoisier and Laplace constructed a 'machine' to study the *calorique* – later termed *calorimètre*. The calorimeter was an instrument to control the combustive process and quantify the development of heat: In the so-called ice-calorimeter, combustible material was shut into a metallic vessel with controlled air supply, which was placed in a larger vessel filled with ice. This second vessel was coated or iced, so that no heat would get lost. The melted ice then provided a measurement of the heat developed in combustion. Calorimeters later came in different forms and designs, but they remained very costly instruments until the invention of Berthelot's bomb calorimeter in 1879 (see Fig. 3.1).[47] Calorimetric analysis did not give results for chemical elements, but specified quantities of 'fixed carbon', 'volatile matter', the water and heat released during the process, as well as its remnants – mostly ash. The calorific value was then calculated from the measured heat and a combination of these quantities. The determination of heating values based on the calorimeter was later called 'proximate' or 'immediate analysis'.

From Pierre Louis Dulong and Alexis-Thérèse Petit's investigations into the heat capacity of solids grew another method to derive heating values from a chemical analysis – a method called 'ultimate analysis'. In 1819, Dulong and Petit presented a law that allowed the calculation of the calorific

45 Robert J. Morris, 'Lavoisier and the Caloric Theory', *British Journal for the History of Science* 6, no. 1 (June 1972): 7–10.

46 McEvoy, 'Continuity and Discontinuity in the Chemical Revolution', 203.

47 Lissa Roberts, 'A Word and the World: The Significance of Naming the Calorimeter', *Isis* 82, no. 2 (1 June 1991): 202–8.

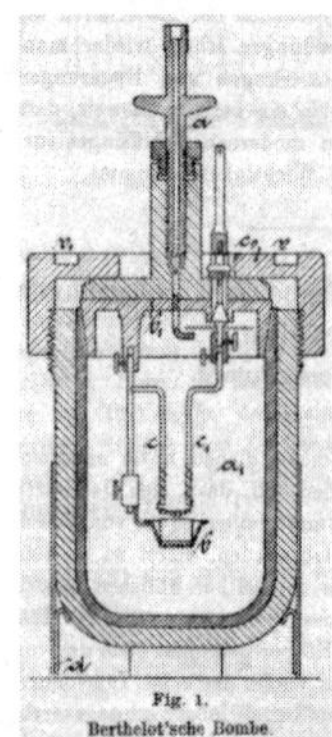

Figure 3.1. Berthelot's calorimeter

value of solid materials based on their chemical composition of the elements – carbon, hydrogen, oxygen, sulphur, and nitrogen.[48] True to the principle of the combination of simple substances, the sum of the calorific value of each element would constitute the overall calorific value of the composed element. Dulong and Petit's law (later referred to as 'Dulong's formula' in the industry) referred to all solid material and was not tailored to the analysis of coal, as they were occupied with atomic theory and the nature of heat. Still, the method was later widely used to determine the calorific value of coals.[49] As a sum of each individual element's heating value in the proportion in which it was present in the coal, the formula (wrongly) assumed that molecules and the composition in which elements were present in an element did not matter.

Coal had long been associated with inflammable minerals, like sulphur, rather than organic matter. However, analysis by Lavoisier, Jöns Jacob Berzelius, and Justus Liebig showed that both the mineral and organic kingdoms were closely related: Not only were there minerals in plants, but the main elements in plants constituted the main elements

48 Pierre Louis Dulong and Alexis-Thérèse Petit, 'Recherches sur quelques points importants de la Théorie de la Chaleur', *Annales de chimie et de physique* 10 (1819): 395–413.

49 Robert Fox, 'The Background to the Discovery of Dulong and Petit's Law', *British Journal for the History of Science* 4, no. 1 (1968): 1–22.

of what had been deemed a mineral – coal. In 1840, J. F. W. Johnston presented an analysis of North British coals before the British Association for the Advancement of Science, arguing that coal was made of vegetable matter and that the different types reflected different states of decomposition.[50] This was later referred to as the 'peat-to-anthracite' theory, which hypothesized that all coals were organic matter exposed to a common geological process. It reconceptualized coal as a single *genus* with *species* reflecting the stages of decomposition. This new classification was based on a geological theory, though its taxonomy reproduced the traditionally noted differences of visible and palpable traits – that is, the way coals felt and broke up, the smoke that was produced, or the length and colour of the flame it produced. Though the species' borders remained indeterminate, the categories established the framework on which geologists and chemists later relied for an emerging classification system. As 'carbon, oxygen, and hydrogen are the component part of living vegetables, and the same element compose coal', it was hoped that the 'true' classification of coal could now be based on ultimate coal analyses.[51] This admittedly ambiguous classification schema was among the most widespread during the nineteenth century, particularly in Great Britain and the US. It survived the changes in coal knowledge of the early twentieth century. In 1932, it was still thought to 'serve [its] purpose of naming' and, today, it constitutes the basis of the ISO's classification system.[52]

Coal had not traditionally been at the centre of chemical and metallurgical analysis, but it began to enter laboratories as a research object in the late eighteenth century. The French and the Prussian Academy of Science announced competitions on the scientific exploration of coal in

50 British Association for the Advancement of Science, 'Tenth Meeting of the British Association for the Advancement of Science, September, 1840', *Journal of the Statistical Society of London* 3, no. 3 (1840): 210.

51 Richard Cowling Taylor and Samuel Stehman Haldeman, *Statistics of Coal: Including Mineral Bituminous Substances Employed in Arts and Manufactures* (J. W. Moore, 1855), 69.

52 Clarence A. Seyler, 'Is Classification or Nomenclature of Coal Possible or Desirable?', *Journal of the Society of Chemical Industry* 51, no. 25 (17 June 1932): 531.

the 1770s.[53] In 1779, the French Ministry of Finance asked Antoine Lavoisier to compare the economic value of fuels available on the Paris market: Apart from prices and taxes, he looked at the amount of fuel needed to evaporate a certain amount of water, a quantity that was later called 'evaporative power'.[54] Decades later, Henri Regnault, a student of organic chemistry with Liebig, was the first to dedicate an article, 'Recherches sur les combustibles minéreaux' (1837), to the chemical study of combustibles. Combustibles have rarely been properly analysed, noted Regnault in the introduction, and if so, they have only been exposed to 'immediate' analysis – to distillation and burning. While this can give an idea of their quality and value, it can never disclose their 'intimate nature' as fuels, their elementary composition.[55] Regnault analysed a wide range of coals and underpinned the geological classification with chemical analyses. When industrialists relied on those analyses, however, they quickly found that the result did not agree with the value of coal in practice.

Universal Steam, Particular Coals

Until well into the nineteenth century, coal was used without much knowledge of what constituted its energetic value, or which properties precisely affected its steam-generating effect. This is due not to a lack of knowledge but to a different knowledge of coal, one not adapted to the needs of the steam economy. As the steam engine established a link between coal fire and mechanical power, it became plausible to describe coal as doing work, and to look for the properties in coal that affected

53 Wendt, *Kohlezeit*, 125–7; see, for the works of Jean Francois Clément Morand and the study of coal in the French Académie, ibid., 184–98.

54 Antoine Laurent Lavoisier, 'Expériences sur l'effet comparé de différentes combustibles', *Mémoires de l'Académie des sciences*, 1781, 377–90. This was no decision against using the calorimeter in the study of coal, as Lavoisier began using it only three years later, in 1782.

55 Henri Victor Regnault, 'Recherches sur les combustibles minéraux', *Annales des mines* 3, no. 12 (1837): 161.

this performance. Manufacturers, industrialists, and scientists had of course tested coals for specific purposes, including their performance in steam engines. Scottish steam engineers had long compared local coals by measuring the work a certain amount of coal could perform in the steam engine.[56] Locally, this method could be used to find the most suitable coals for a given engine or to measure the efficiency of the engine against the same type of coals. These tests simply related a measure of weight to a measure of work, without an understanding of what conditioned that relationship. The process between the ignition of coal and the rotation of the wheel remained a black box; it was impossible to know how much steam was raised or how much heat produced at the time.[57] Even if the Scottish steam engineer's test results had made their way beyond a particular mine, mill, or region, they would be of little value elsewhere as they did not allow to distinguish between the effects of the coal and those of the experimental apparatus.

Meanwhile, coals of the world's largest mining regions began to reach more distant places. In the first half of the nineteenth century, metallurgy, steam engines, and coal began to form a mutually reinforcing pattern.[58] From then on, coal was capable of powering and moving itself across the globe. Railways and steamships, forged with cheap iron fuelled by coal and financed by the regular revenue of coal shipments, extended the area over which coal could be transported.[59] British coal travelled

56 Donald Fleming, 'Latent Heat and the Invention of the Watt Engine', *Isis* 43, no. 1 (1 April 1952): 4.

57 The steam engine could be fitted to the productive apparatus only by trial and error until engines featured a steam indicator and a diagram in the 1790s; R. L. Hills and A. J. Pacey, 'The Measurement of Power in Early Steam-Driven Textile Mills', *Technology and Culture* 13, no. 1 (January 1972): 25. Even then, coal remained outside this calculatory relationship for another half century.

58 Astrid Kander, Paolo Malanima, and Paul Warde, *Power to the People: Energy in Europe over the Last Five Centuries* (Princeton University Press, 2014), 92–3; Brüggemeier, *Grubengold*, 94; Alessandro Nuvolari, Bart Verspagen, and Nick von Tunzelmann, 'The Early Diffusion of the Steam Engine in Britain, 1700–1800: A Reappraisal', *Cliometrica* 5, no. 3 (October 2011): 291–321.

59 Freese, *Coal: A Human History*, 61–3; Kander, Malanima, and Warde, *Power to the People*, 192–3; John E. Murray and Javier Silvestre, 'Integration in European Coal Markets, 1833–1913', *Economic History Review* 73, no. 3 (2020): 668–702.

most easily, since the coal fields were close to ports and transport over sea was by far the cheapest. It was traded globally already by the early nineteenth century, and the urban centres of the North American seaboard consumed British coal in the 1820s, as did Italy, the wider Mediterranean, and the Russian Empire.[60] As cargo or ballast on the empire's steam and sailing fleet, British coal could reach nearly every corner of the world.[61] The other two large nineteenth-century coal producers, the United States and Germany, had to haul their coal over land, and this limited the area it could reach at cost-effective prices.[62]

The introduction of steam technology carried the search for fuel to faraway places and generalized it, but it did not at first require an *energetic determination* of those fuels. The uptake of steam technology did not coincide with the uptake of coal everywhere: If only fossil-fuelled fire could have yielded steam, its adoption would have been limited to places that coal could reach. That steam can be generated from heating water over a fire of *any* fuel, fossil or not, facilitates its spatial and temporal independence. The early diffusion of the steam engine rested on this use of cheaply available fuels, often biomass. In the Caribbean sugar industry, steam engines burned scraps from agriculture; where coal was distant or

60 Alfred D. Chandler, 'Anthracite Coal and the Beginnings of the Industrial Revolution in the United States', *Business History Review* 46, no. 2 (June 1972): 151; N. V. Brivko, 'Stanovlenie ugol'noi promyshlennosti konca XIX veka–nachala XX veka (na primere goroda Snezhnoe Doneckoi oblasti)', *Istorichnyj archiv* 12 (2014): 35.

61 On Barak, 'Outsourcing: Energy and Empire in the Age of Coal, 1820–1911', *International Journal of Middle East Studies* 47, no. 3 (August 2015): 427.

62 It took the financing and construction of canals and railways to mobilize landlocked Pennsylvanian coal and make it cheap enough for American river steamboats to replace wood as a fuel; Frederick Moore Binder, 'Pennsylvania Coal and the Beginnings of American Steam Navigation', *Pennsylvania Magazine of History and Biography* 83, no. 4 (1959): 420–5; Chandler, 'Anthracite Coal and the Beginnings of the Industrial Revolution'; Christopher F. Jones, 'A Landscape of Energy Abundance: Anthracite Coal Canals and the Roots of American Fossil Fuel Dependence, 1820–1860', *Environmental History* 15, no. 3 (2010): 449–84. German coal producers faced similar problems: In a city as close to the German deposits of Hamburg, British coal could easily compete with Ruhr coal transported by railway; A. J. Sargent, *Coal in International Trade* (P. S. King, 1922), 25–6.

wood abundant, it could make sense to burn down forests.[63] In some places, the use of wood to raise steam continued until far into the nineteenth century.[64] But even where coal was used for steam generation, it could be done so for a long time without much knowledge on coal's evaporative power – and certainly without knowledge that applied to *all coals*. The practical use of fuels for steam generation predates their general energetic valuation.

Railways and steamships enlarged coal markets. This meant that consumers now could choose between more different coals – and that they had to do so. At the same time, coal-fired railways and steamships emerged as large coal consumers who bought massive amounts of coal under long-term contracts. Their service relied on self-moving steam engines that needed to carry their fuel with them. For those consumers, even small changes in the quality of coal mattered and analysis assumed a more important role. By the mid-nineteenth century, coal analysis was linked to a reduction of transport costs explicitly. 'It is obvious that the price of coal is considerably affected by the transport of the barren parts included in it', noted Ernst Hartig, a Saxon engineer who had been trained in a company that constructed steam engines for the Royal Saxon railroads, in 1865.[65] The link between transport costs and weight generated an imperative to re-examine coal's value.

Close study of fuels resulted in a growing distinction between those properties that enhanced and those that impeded combustibility. That there was a varying non-combustible part in coal had been known before – ash in fireplaces and furnaces attested to that. There was, additionally, sulphur, an element known to occur in coals, and which could make them

63 Jennifer Tann, 'Steam and Sugar: The Diffusion of the Stationary Steam Engine to the Caribbean Sugar Industry 1770–1840', *History of Technology* 19 (1997): 77; David A. Tillman, *Wood as an Energy Resource* (Elsevier, 2012), 6.

64 Peter A. Shulman, *Coal and Empire: The Birth of Energy Security in Industrial America* (Johns Hopkins University Press, 2015), 43–6; Clive Dewey, *Steamboats on the Indus: The Limits of Western Technological Superiority in South Asia*, 1st ed. (Oxford University Press, 2014).

65 Ernst Hartig, 'Die Aufbereitung der Steinkohlen', in Hugo Fleck and Ernst Hartig, eds, *Geschichte, Statistik, und Technik der Steinkohlen Deutschland's und anderer Länder Europas* (R. Oldenbourg, 1865), 332.

unsuitable for metallurgical purposes. Dulong's formula, however, counted sulphur as an element conducive to heating and the formula entirely disregarded ash. What might be neutral from the chemical point of view of combustion, however, was not neutral from the commercial perspective, as unproductive elements, like ash, added to the weight and volume of coal, but not to the effect it yielded. Abstract valuation is total, it permeates the entire rock, and leaves no space for neutral components.

Against this backdrop it becomes understandable that the first large-scale industrial tests of coals for steam navigation were conducted by navies. Ocean steam navigation was one field where the quantity of steam generation and the power of engines mattered from the beginning. Wood could never power ocean steamers: They required a fuel that did not take up too much space and could power the vessel over long distances.[66] Steamships were introduced to the British Navy in the early nineteenth century (they remained hybrids until 1871).[67] Not only could those ships be made of heavier materials that could withstand attacks, they also ended the Navy's dependence on wind and current, and enabled direct and calculable routes and higher speed.[68] Until the end of the century, steamships remained mainly limited to the military, mail, and passenger lines; cheap bulk products were still overwhelmingly transported by sail.

To find the suitable coals and legitimize their purchases from certain coal areas, several admiralties commissioned large trials of coal in the mid-nineteenth century.[69] They found that chemical analyses alone

66 Gerald S. Graham, 'The Ascendancy of the Sailing Ship 1850–85', *Economic History Review* 9, no. 1 (1956): 74–88.

67 Steven Gray, 'Black Diamonds: Coal, the Royal Navy, and British Imperial Coaling Stations, Circa 1870–1914' (PhD thesis, University of Warwick, 2014), 144; Crosbie Smith, *The Science of Energy: A Cultural History of Energy Physics in Victorian Britain* (University of Chicago Press, 1998), 34–5; Crosbie Smith, *Coal, Steam and Ships: Engineering, Enterprise and Empire on the Nineteenth-Century Seas* (Cambridge University Press, 2018), 172–3.

68 Steven Gray, 'Fuelling Mobility: Coal and Britain's Naval Power, c. 1870–1914', *Journal of Historical Geography* 58 (October 2017): 24.

69 Gray, 'Black Diamonds', 141–4; Frederick Moore Binder, *Coal Age Empire: Pennsylvania Coal and Its Utilization to 1860* (Pennsylvania Historical and Museum Commission, 1974), 13.

could not predict the performance of coal to a degree of precision needed in practice. Therefore, engineers tested coals in an experimental setting that mirrored their practical application as closely as possible. This meant to burn not only small samples, as chemists did, but hundreds, or even thousands, of pounds of coal under an industrial- or marine-style boiler. At the same time, by relating their assays to each other, they began to standardize the experimental settings to test steam-raising and were looking for recurring relations between chemical analysis and practical performance under the testing boiler.

Welsh steam coal was the fuel of choice for ocean navigation and powered not only the British but most other marines around the world in the first half of the nineteenth century. In 1842, wary of the US Navy's dependence on foreign coal, the US admiralty commanded the first systematic testing of a nation's coal. The purpose was to 'increase the efficiency of the navy' vis-à-vis the British. Over two years, Walter Johnson tested forty-five American coals mainly with a focus on their suitability for steam navigation.[70] Using proximate analysis next to assays of evaporative power, he arranged coals according to their relation of fixed carbon to volatile matter, trying to show that this 'fuel-ratio' had a constant relation to the evaporative power of the tested fuels. This ratio can be seen as the first general 'index of coal values'.[71] Though not picked up at the time, the fuel-ratio was an attempt to establish the missing link in the arithmetic connection between the effect of motive power, the properties of steam, and the elements of coal.

Johnson's report sparked similar efforts in Great Britain, Prussia, and other countries.[72] The British admiralty ordered Sir Henry de la Beche

70 Navy Department of the United States and Walter Johnson, 'American Coals', 1844, 47–9.

71 Samuel W. Parr, 'The Classification of Coal', *University of Illinois Bulletin* 25, no. 48 (31 July 1928): 6–7.

72 In Saxony, the government commissioned a three-part study on coal in the 1850s, covering geological and chemical questions, as well as heating values of 139 Saxonian coals (the latter two volumes are Stein's *Chemische und chemisch-technische Untersuchungen der Steinkohlen Sachsens* [Chemical and chemical-technical investigations of Saxon coals] from 1857 and Ernst Hartig's *Untersuchungen über die Heizkraft der Steinkohlen Sachsens* [Analyses on the heating value of Saxon coals]

and Lyon Playfair to ascertain 'what fuels have the greatest evaporating power in the smallest space and weight', as 'without an accurate knowledge of the power of the coals . . . the public service may suffer . . . when the greatest interests of the country may be at stake'.[73] Just like Johnson, de la Beche and Playfair emphasized the gap between laboratory analyses and a coal's value in practice. While they performed chemical analyses on each coal, they also tested the evaporative power practically and 'on a scale of sufficient magnitude' to compare theoretical and practical results.[74] The British admiralty alone ordered over twenty trials of domestic and foreign coals between 1847 and 1879.

In Prussia, Johnson's report met with the aspiration held by influential Prussian industrialists to emulate British industrialization. It prompted the head of the Prussian Association for the Promotion of Technical Activity (Verein zur Beförderung des Gewerbefleißes), Peter Beuth, to fund a similar investigation almost single-handedly in Prussia in 1848. Beuth's mission in founding the association had been to arm the Prussian industry so it could weather world competition: 'If a state abandons its factory owners . . . exposes them to the wind and weather of competition – like the Prussian does – it has the obligation to make them familiar with the means by which they can survive competition.'[75] For Beuth, the energetic determination of coal had, by then, become a necessary means for all kinds of industries to succeed on the world market.

The cost and scope of such large-scale testing of nationally available coals were beyond the means of single companies. It had to be done by

---

from 1860). An analysis of Austrian coals follows in 1862; a year later, the Royal Prussian Navy's main shipyard in Danzig compared English, Prussian, and Saxon coals for steam navigation.

73 Henry T. de la Beche and Lyon Playfair, *First [-Third] Report on the Coals Suited to the Steam Navy* (W. Clowes and Sons for HM Statistical Office, 1848), 6.

74 De la Beche and Playfair included specimens of two objects of British colonial ambition – Borneo and Formosa – in their analyses. But they were not sure how well the long-travelled samples represented the deposit and were unable to get hold of the tons of coal needed for testing their evaporative power.

75 Helmut Reihlen, *Christian Peter Wilhelm Beuth: Eine Betrachtung zur preußischen Politik der Gewerbeförderung in der ersten Hälfte des 19. Jahrhunderts und zu den Drakeschen Beuth-Reliefs* (Beuth, 1992), 21–2.

an institution that was capable to erect a laboratory, acquire a testing boiler, equipment, and measuring devices, and purchase scientific labour for at least a year – usually longer. With government funding, engineers, chemists, and geologists from North America's and Europe's coal regions began to develop classifications based on what was now considered the common denominator of all industrial uses – the heating (or calorific) value.[76] While the heating value was nowhere near as important as in steam generation and transport, it was useful in steelmaking as well (when accompanied with other analyses). Some authors of the treatises were themselves industrialists, and all were committed to the development of regional natural resources, but they no longer tested coal for a specific or isolated task. These were books for specialists, with detailed descriptions of method, but they aimed for the general perspective of industry.[77] This knowledge was to be general enough, at least, to be applicable to all fuels in a given region.[78] Whether classification could ever be transregional was contested: As late as 1887, German coal chemist Friedrich Muck – another student of Liebig who had turned to the analysis of coal – was convinced that classification across coal districts would be impossible to achieve and that schemes would have to remain

76 Ernst Hartig, 'Die Leistung der Steinkohlen als Brennstoff', in Fleck and Hartig, *Geschichte, Statistik, und Technik der Steinkohlen Deutschland's und anderer Länder Europas*, 289. See, for published coal analyses around the time, Richard C. Taylor's *Statistics of Coal* (1848), Pierre Berthier's *Traité des essais par la voie sèche des propriétés, de la composition, et de l'essai des substances metalliques et des combustibles* (1848), Wilhelm Brix's *Untersuchungen über die Heizkraft der wichtigeren Brennstoffe des preussischen Staates* (1853), Muck's *Chemische Aphorismen über Steinkohlen* (1873), Gruner's *Traité de métallurgie* (1878), and Auguste Scheurer-Kestner's *Pouvoir calorifique de combustibles solides, liquides, et gazeuse* (1896).

77 Nora Thorade, *Das schwarze Gold: Eine Stoffgeschichte der Steinkohle im 19. Jahrhundert* (Ferdinand Schöningh, 2020), 195.

78 Louis Gruner's classification system, which was a refinement of Regnault's mentioned above, brought together aesthetic qualities of the coal (its texture and the length of its flame) with elementary analysis, and was widely used in Europe; Louis Gruner, *Traité de métallurgie*, vol. 1 (Dunod, 1878). In the US, in turn, a system based on ultimate analysis along Johnson's fuel-ratio was developed. In general, such classification systems rested on only a dozen to a hundred different coals – those deemed relevant for a given region; Thorade, *Das schwarze Gold*, 100.

regional.[79] Fifty years later, some national classifications were still deemed entirely 'meaningless' for the foreigner.[80]

At the same time, existing coal analyses and classifications could not fully satisfy commercial interests. In absence of a standard method, the books gave long, meandering descriptions of the procedures they followed and the devices they used. Replicating results was almost impossible and divergent findings could have a myriad of reasons other than a real difference in the coal tested, such as differences in sampling, storage, climate, instruments, or procedure. Comprehensive coal testing took years, while markets changed in terms of weeks or months, and 'no one would claim that coal of one and the same provenance can be supplied of quite the same quality for only a few months'.[81] Testing was doomed to a retrospective determination of value, but what was really needed was a method to predict a coal's performance from a simple test that could be performed on a small sample. Wilhelm Brix, who had conducted the Prussian coal tests, saw clearly what was missing: a systematic study that would combine a test of steam generation with calorimetric (proximate) and chemical (ultimate) analyses. On this basis, it 'might be possible to develop a method from which the heat effect of a material can be deduced from its composition'.[82] Once all coals were analysed and sorted in that way, their value could be derived from a classification and trade could finally be based on the 'value-determining characteristics' of the raw material.[83] In a world in which industrialists had access to a variety of global coals and could gain an

79 F. Muck, *Elementarbuch der Steinkohlen-Chemie für Praktiker* (G. D. Baedecker, 1887), 25–6.

80 Parr, 'The Classification of Coal', *University of Illinois Bulletin*, 6.

81 Franz Schwackhöfer, 'Calorimetrische Werthbestimmung der Brennmaterialien', *Zeitschrift für analytische Chemie* 23, no. 1 (1 December 1884): 456.

82 Wilhelm Brix, *Untersuchungen über die Heizkraft der wichtigeren Brennstoffe des preussischen Staates: Im Auftrage des Vereins zur Beförderung des Gewerbefleißes in Preußen und mit Unterstützung des Königlichen Ministeriums für Handel und Gewerbe* (Ernst & Korn, 1853), 6.

83 O. Mohr, 'Die Analyse als Grundlage für die Kohlenbewertung und den Kohlenhandel', *Angewandte Chemie* 21, no. 40 (2 October 1908): 2089.

advantage by choosing the right one, a theoretical inference of the heating value would indeed be very practical.

The hope to pin down coal's real value by systematically comparing the different methods of analysis – elementary, proximate analysis, and tests of steam-raising – was quickly disappointed. In 1868, Auguste Scheurer-Kestner, an Alsatian politician, chemist, and entrepreneur, had shown that Dulong's formula to calculate the heating value from chemical elements did not coincide with heating values determined by the calorimeter.[84] Over the following decades, studies found a variation between the methods of 1 and 10 per cent depending on the coal tested. Elementary analysis – the more 'intimate' method – confusingly realized lower values. This was a meagre result for a method that had by then become the official formula of the German Association for the Testing of Steam Boilers (Dampfkesselüberwachungsverein) and raised the question whether 'the way in which the elements of a hard coal are connected with each other is too different, too fluctuating, for the thermal effect of it to be deduced from the results of the simple elementary analysis'.[85] In a large coal study at the Munich Heat Testing Station (Heizversuchsstation), Hans Bunte sought to replicate Scheurer-Kestner's results, but found that differences were much less stark than expected and that elementary analysis could well serve as a rough indication of the heating value. A rough indication, however, was no longer enough for industrial interests. By then, the industrial testing of heating values had already shifted to the calorimeter. Not only was it considered more accurate for the measurement of heat, but it also corresponded to the commercial need to perform analyses more regularly, cheaply, and quickly. By the late nineteenth century, several chemists had developed new calorimeters that were affordable, easier to handle, and made it possible to conduct analyses by consumers themselves.

84 M. A. Scheurer-Kestner, 'Recherches sur la combustion de la houille', *Bulletin de la Société Industrielle de Mulhouse*, April–May 1868, 195–251.

85 L. Gruner, 'Über die Heizkraft und die Classification der Steinkohle', *Polytechnisches Journal* 213, no. 11 (1874): 70.

The Idea of Fuel

Confusing as it may have been, the study of coal's elementary composition and its relation to the heating value enabled the objectification of fuel on a higher level. This was the programme of fuel science, a research field that emerged in the interwar years.[86] Coal's value may no longer be visible or palpable, but if something like a coal molecule could be found, if there was a hidden order in the relations between elements that were present in coal, not all hope would be lost. If this order had a relation to the heating value, and could be represented, it would open a new global plane of comparison. As the British coal chemist Clarence A. Seyler noted, the 'real identity' and 'real diversity' of coals lay hidden unless brought to light by scientific analysis.[87] In the early twentieth century, fuel became a hypothetical substance, one that – like phlogiston – could not be realized in the laboratory. What scientists called 'pure coal substance' or 'true coal' could never be fixed in crucibles or test tubes: It was the result of a calculation that derived its meaning from the steam industry's need of an obediently burning fuel.

The multiplicity of coal classifications and methods of analysis was a problem for those economic interests that transcended regions and sought control over a national or global topography of fuel.[88] Large consumers pushed for coal analyses to become the basis of coal trade. With more efficient engines, turbines, and furnaces, and more sophisticated business models to navigate competition in an industry in which steam power had long become a necessity, small variances in the quality of coal could matter. Apart from the railroads, steamers, and large manufacturing combines, electric power plants emerged as large consumers in the late nineteenth century. As we will see in the next chapter, power

86 Research along these lines was published in the new journals *Brennstoffchemie* (1920), *Fuel in Science and Practice* (1922) (as a supplement to the *Colliery Guardian*), *Fuel Economist* (1925), and *Fuel Economy Review* (1921).

87 Seyler, 'Classification of Coals and the Interpretation of Analysis', vii.

88 World Power Conference, ed., *The Transactions of the Fuel Conference: World Power Conference, London, September 24–October 6, 1928* (P. Lund, Humphries & Co., 1929), 257.

> this amount is expended in the collection and disposal of the ashes from the place of consumption to their final resting place . . . The Country's non-productive expenditure on this score may, therefore, be safely put at the stupendous figure of from £10–20,000,000, a waste which surely should be a worthy subject of the closest scientific, technical and economic examination.[94]

Examination of coal was already under way, as coal research institutes popped up around the world. The US Geological Survey had already set up its Coal Testing Station in 1904, 'to undertake a program of analyzing and testing coals and lignite to determine their fuel value and the most economical methods for their utilization'.[95] At the University of Illinois, the engineering experimental stations set out 'to study problems of importance to professional engineers and to the manufacturing, railway, mining and industrial interests of the United States'.[96] Some years later, the Geological Survey merged with the US Bureau of Mines, a new apparatus to oversee efficiency in the industry, which included the valuation of coal lands that the state would lease for exploitation.[97] Across the Atlantic, the coal and power industry of the Ruhr area cooperated with Prussian officials to found the Kaiser-Wilhelm-Institut für Kohlenforschung in 1913 – an institute to study the nation's 'treasures of power and feedstock'.[98] Not everywhere was the link between fossil capital and the state as strong as here, where public money and the profits from the mining, power, and metallurgical

94 Richard Lessing, 'Coal Ash and Clean Coal', *Journal of the Royal Societies of Arts* 1 (1926): 183.

95 Mary C. Rabbitt, *A Brief History of the U.S. Geological Survey* (US Department of the Interior/Geological Survey, 1974), 19.

96 Samuel W. Parr and Wilfred F. Wheeler, preface to *Unit Coal and the Composition of Coal Ash* (University of Urbana-Champaign Press, 1909).

97 W. R. Calvert, 'Land Classification, Its Basis and Method', *Economic Geology* 6 (1911): 478; Fred Wilbur Powell, *The Bureau of Mines, Its History, Activities and Organization* (D. Appleton and Company, 1922).

98 K. H. van Heek, 'Progress of Coal Science in the 20th Century', *Fuel* 79, no. 1 (January 2000): 4.

industries jointly funded coal research.[99] In the UK, more targeted state-funded research on fuel began in the middle of World War I with the foundation of the Department of Scientific and Industrial Research.[100] In Japan, fuel research was institutionalized a few years later.[101]

It was hoped that coal trade would come to rest more and more on the 'value-bearing' characteristics of coal, primarily the calorific value.[102] Apart from volume and price, contracts now often stipulated a range in which the calorific value and other properties were allowed to vary. Large combines and corporations could even establish their own buying department with a technical staff. 'These technical experts, acting on behalf of their company, have evolved a system of testing and evaluating . . . and have formed for themselves standards by which all [coal] supplies are graded, depending upon the purpose for which the coal is required' – a procedure called 'scientific buying'.[103] Those large consumers could use their market power to force sale under a guarantee.[104] Despite sorting, washing, and processing the coal, coal companies were not able to guarantee a stable product and had to put up with price reductions when the buyer could prove a declining quality.[105] As late as the 1920s, only a few collieries could 'furnish reliable and authoritative analyses of the fuels they [were] offering', and complaints about a lack of

99 Manfred Rasch, *Geschichte des Kaiser-Wilhelm-Instituts für Kohlenforschung 1913–1943* (VCH Verlagsgesellschaft, 1989), 21–32.

100 Wilson, *Down to Earth*, 38.

101 Victor Seow, '"Coal Will Be the Primary Fuel of the Future": Yoshimura Manji on the "Fuel Question"', in Daniela Russ and Thomas Turnbull, eds, *Energy's History: Toward a Global Canon* (Stanford University Press, 2025), 37.

102 Mohr, 'Die Analyse als Grundlage', 2089; F. Haber and S. Grinberg, 'Ueber Elementaranalyse von Kohlen', *Fresenius' Zeitschrift für analytische Chemie* 36, no. 1 (December 1897): 557.

103 R. A. Burrows and N. Simpkin, 'The Sale of Coal Under Specification', in World Power Conference, ed., *The Transactions of the Second World Power Conference*, vol. 6, sec. 12 (VDI Verlag, 1930), 322.

104 Mohr, 'Die Analyse als Grundlage', 2094.

105 Schwackhöfer, 'Calorimetrische Werthbestimmung der Brennmaterialien', 456–7.

standards of analysis, the representation and interpretation of results, and coal classification were prevalent.[106]

In regional markets, this insecurity of value had been addressed by choosing one particular coal that would serve as a point of reference for all others. Not every coal was suitable for the role. A reference coal had to be well known and widely accessible, and relatively stable in quality. In India, for instance, coal from the Karharbari field had been used as the benchmark for all railway fuel, even timber, since the 1870s.[107] British collieries used complex constructions to relate their coals to each other, while the American Anthracite Companies agreed on the 'white ash coal' to serve as an industry standard according to which all other coals were priced.[108] While these standard coals were rooted in an existing coal, they were already abstracted from the particular forms in which that coal existed: They were fixed as an average, or according to their ash- and water-free basis. 'This view concedes', wrote the American coal scientists Samuel Parr and Wilfred Wheeler in 1909, 'that coal from a certain locality or seam does not vary in quality, but that the variation is due to the presence of ash and moisture which are impurities associated with the coal.'[109] Such standard coals were first introduced only as regionally or purpose-specific constants of calorific value, with no claim to universality.[110]

In the first decades of the twentieth century, this idea of a reference coal was picked up by coal scientists in the US and Great Britain in their attempt to find a global standard: a coal substance that all coals shared and on which global classification could be based. However, this hidden

106 World Power Conference, *The Transactions of the Fuel Conference: World Power Conference, London, September 24–October 6, 1928*, 257.

107 Matthew Shutzer, 'India's Political Ecology of Coal, 1870–1975', unpublished manuscript, 2018, 11; Matthew Shutzer, *Until the Last Ton: Fossil Fuels in India from Empire to the Climate Crisis* (Princeton University Press, forthcoming).

108 Seyler, 'Classification of Coals', 13; Frederick E. Saward, *The Coal Trade: A Compendium of Valuable Information* (111 Broadway, New York, 1875), 7; Binder, *Coal Age Empire*.

109 Parr and Wheeler, *Unit Coal and the Composition of Coal Ash*, 4.

110 Ibid., 3; World Power Conference, *The Transactions of the Fuel Conference*, 271.

identity would become visible only once analysis was based no longer on actually existing coals but on the non-accidental, combustible parts of coal, which were variously called 'actual', 'true', or 'pure coal'. The results of analyses appear unintelligible, warned Clarence A. Seyler, 'as long as they are complicated by the inclusion of matters which are *not coal*, either water or mineral matters'. This is not to say that the amounts of water, ash, and sulphur could be ignored; they were important to the coal user. 'But for the purpose of comparing the composition of the true coal they are accidental and unessential and must be eliminated before any rational comparison can be made.'[111] Similarly, Parr echoed that coal analysis, and especially the study of heating values, would be helped by focusing on the 'properties of the pure coal substance as distinct from the non-coal material'.[112]

Proximate analysis had directed attention to the substance within a coal that resisted combustion, and this substance had been found to consist primarily of mineral matter (often simply called ash). Still, the 'non-coal' that Parr and Seyler had in mind did not fully cohere with that residue. By the early twentieth century, it was clear that non-combustible substances were pervasive in coal: Minerals and moisture were associated with the combustible material not as water to a sponge, but as chemical compounds, or molecules. Worse still, this inert non-combustible matter tended to form new compounds as it reacted with the air during combustion. Traditional proximate analysis therefore counted these substances as volatile matter and included them in the calculation of calorific values.[113] Seyler and Parr enjoined analysts to cease treating the material with which they worked as coal. The true coal substance was neither encountered in nature, nor realized as a result of chemical analysis. It was a *differential* value: True coal was coal minus a quantity recalculated according to the combined oxygen, minerals, and sulphur – all elements deemed useless in production.

111 Seyler, 'Classification of Coals', vii.

112 Parr and Wheeler, *Unit Coal and the Composition of Coal Ash*, 2.

113 Samuel W. Parr, 'The Classification of Coal', *Journal of Industrial and Engineering Chemistry* 14, no. 10 (1 October 1922): 28–30; Seyler, 'Classification of Coals', vii.

Unless classification was based on this substance that was isolated by calculation, it would fail to be comprehended as coal as such. Seyler provided an example of why this had to be so: When classification referred to a sample that included 'impurities', 'identical' coals would be sorted into different classes. Anthracite of the same rank, but with a different amount of ash, at levels of 1 per cent and 10 per cent, would be classified very differently. 'Suppose the anthracite in the pure state to contain 93 per cent of carbon', Seyler explained, 'then in the first case the carbon will be reduced to 92 per cent, and in the latter to 83.7 per cent. The coals will appear to be entirely different, whereas elimination of the ash will show them to be identical.'[114] In this way, the method could unearth the similarity of coals that were formed far from each other and that were hitherto thought to be different. Seyler and Parr stressed the scientific advantages of this approach, as it allowed analysts to focus on the study of 'the main phenomenon' in coal – 'development' – and to base classification upon it.[115]

What coal chemists called 'impurities' or 'non-coal elements' were not only elements that occurred naturally in coal but elements that were to a certain extent present in *all* naturally occurring coals. The difference is not between a natural process of coal formation and an unnatural association with mineral matter, but rather between development and accident. Parr described the non-coal material as 'extraneous or adventitious material which by accident or through natural causes may have become associated with the combustible organic substance of coal'.[116] Not everyone accepted water as an 'extraneous' part of the pure coal substance and the coalification process. Water, minerals, and sulphur all play a role in plant organisms and are thus not as 'accidentally' present in coal as Parr claimed. Some saw water as a 'normal and characteristic coal component'. Ash and sulphur, however, were generally regarded as 'more or less fortuitous' – not because they lacked standing as natural

114 Seyler, 'Classification of Coals', viii.

115 Ibid., 532.

116 Samuel W. Parr, 'The Classification of Coals', *Journal of the American Chemical Society* 28, no. 10 (1906): 1426–7.

components of coal, but because their occurrence followed no conceivable developmental logic.[117] Coal now referred only to the substances in coal that were subject to a logical development.

Even reduced to their 'true' coal substances, however, globally occurring coals varied in multiple dimensions. No single quantity of proximate analysis, and no one chemical element, could account for the development from peat to anthracite. The relation of fixed carbon to volatile matter, the fuel-ratio, worked well for coals on the bituminous-anthracite spectrum. But as European scientists had long pointed out, such a system 'failed egregiously' for lignite and peat.[118] Because proximate analysis did not allow for the separation of oxygen found in coal from the oxygen 'escaping' from it as volatile matter, it failed to register the differences between lower-rank coals. To differentiate within lower-rank coal, moisture was crucial. In Europe, Gruner's classification based on the proportion of oxygen was more widespread, and it could draw finer distinctions for the lower-grade coals, but only modest ones for other types. For that reason, several scientists experimented with two-dimensional classification systems at that time – one from peat to brown coal, and one from bituminous to anthracite.[119]

The identification of a pure coal substance was the objectification and projection of coal's capacity to do work onto the coal material. This substance does not necessarily coincide with the maximum heating value but rather with the maximum regular work. The pure coal substance allowed scientists to develop methods of analyses from which the heating value could be derived within a certain margin of error, but this did not mean that the water-, ash-, and sulphur-free pure coal substance indicated the upper limit of the heating value that could be calculated or achieved in practice. It is true that water and ash reduce the heating value, as the former uses up part of the heat for evaporation and

117 D. J. Fisher, 'Notes Regarding the Coalification Process', *Journal of Geology* 35, no. 7 (1927): 639–46.

118 Seyler, 'Classification of Coals', x.

119 Clarence A. Seyler, for instance, published his classification system based on hydrogen and volatile matter already in 1900; Clarence A. Seyler, 'The Chemical Classification of Coal', *Fuel in Science and Practice* 2 (1900): 272–3.

the latter can be seen as idle substance. Sulphur, however, was an entirely different matter: It could affect the heating value positively but was also associated with the recalcitrant behaviour of some coals. In the form of iron pyrites, it was said to cause spontaneous combustion. Its positive effect on the heating value remained too unruly, its heating activity too uncontrollable to be considered part of the fuel substance. The pure coal substance thus limited coal to the docile parts contributing to a controllable combustion.

From a more practical perspective, this approach outlined an idealized conception of a fuel. Parr's and Seyler's pure coal substance was part of every coal but not identical with any coal. The 'pure coal substance' constituted an 'individual' ideal for each type of coal. Recalculating the measured quantities as if they were based solely on the valuable part of coal meant classifying coals according to the best coal they could be. Bituminous coals would never become anthracite, but they might well become a particularly pure bituminous coal. The 'pure coal substance' was a thing that could be handled but had no physical existence. As a hypothetical coal, however, it was as formal and well defined as any mathematical object and opened new possibilities for refinement, to the end of increasing coal's use value. Parr and Seyler understood scientific analysis as a means of revealing the true identity of coal, undetectable by simple observation and beyond the scope of the old methods. This knowledge revealed unexploited commercial potential. New means for modifying coal could now be mobilized to enhance its production and the profit drawn from it.

Looking back from the Fuel Conference in London in 1928, Seyler was struck by the fundamental similarities between his classification system and Parr's. Even though they had employed two distinct methods of analysis, proximate and ultimate, they nonetheless ended up sorting coals into similar classes – a powerful indication that a true, hidden order had been discovered.[120] Experiments with coal's molecular structure added to this conviction: From 1910 on, Friedrich Bergius

120 World Power Conference, *The Transactions of the Fuel Conference*, 254.

developed a process for treating coal under high pressure which emulated the creation of stone coal. In effect, Bergius transformed peat into a soft coal resembling the bituminous variety. A coal of 84 per cent carbon constituted his experiment's stable end-product, and this fact appeared to confirm that two processes were at work in coal's formation: one moving from peat to bituminous and another from bituminous to those with an even higher carbon content, such as anthracite.[121] Yet this geological insight was itself only a by-product of what Bergius really sought – a synthetic liquid or gaseous fuel derived from coal. Bergius's liquefaction of coal, later applied at a larger scale in I. G. Farben's synthetic fuels programme, was one of many experiments at that time to transform coal into a homogeneous and stable fuel. Others attempted to produce a 'constant fuel element' by circumventing combustion and generating electricity directly from coal.[122]

This view of coal was shaped by the rise of petroleum, a more concentrated and homogeneous fuel, which could replace coal for certain applications (notably navigation, transport, or stationary industrial engines). In the geopolitically fragmented market of the early twentieth century, where not all states had access to oil resources, this comparison led not only to coal substitution, but also to the opposite – a more intensive use and processing of all types of coal. In the name of economic autarky, fuel scientists and chemists remade a century-old fuel to suit this new energy economy. Coal, power, and petroleum industries grew to be tightly

121 Anthony N. Stranges, 'Friedrich Bergius and the Rise of the German Synthetic Fuel Industry', *Isis* 75, no. 4 (1 December 1984): 651–2.

122 This was the problem Franz Fischer set himself to solve as head of the Kaiser-Wilhelm-Institut for Coal Research; Rasch, *Geschichte des Kaiser-Wilhelm-Instituts für Kohlenforschung*, 49. Several countries set up synthetic fuel programmes at the time; Anthony N. Stranges, 'From Birmingham to Billingham: High-Pressure Coal Hydrogenation in Great Britain', *Technology and Culture* 26, no. 4 (October 1985): 726; Anthony N. Stranges, 'Canada's Mines Branch and Its Synthetic Fuel Program for Energy Independence', *Technology and Culture* 32, no. 3 (July 1991): 521; Anthony N. Stranges, 'The US Bureau of Mines's Synthetic Fuel Programme, 1920–1950s: German Connections and American Advances', *Annals of Science* 54, no. 1 (January 1997): 29–68; Victor Seow, *Carbon Technocracy: Energy Regimes in Modern East Asia* (University of Chicago Press, 2021), 201–2, 234.

interconnected, and coal-based chemistry came to demand electricity at a colossal scale. Coal's by-products, in turn, would furthermore be used to refine petroleum.

As the history of coal use spans the pre-industrial and industrial eras, we can study how a material became energetically determined and how coal, a highly complex and locally diverse material, resisted behaving as an energy commodity. Steam-powered industrial capitalism drove the energetic valuation of fuels: The combustible substance is no longer the bearer of a principle of fire but a source of a quantifiable amount of heat and work. In this process, chemists reified coal, fuel, and combustibles into a hypothetical substance, unrealized in nature, and made up purely of elements that took part in the natural process of 'fuel development'. To coal scientists, this brought to light a hidden order in the confusing variety of globally occurring coals. With this concept, every coal in existence could be related to a standard one, made up of purely combustible material. Yet this was not a purely descriptive order: It charged coal with potentialities, outlined its more extensive and intensive use, and opened a space for its technical development into synthetic fuels. From the beginning, it also served as an ideal towards which coals could be developed. Once pricing followed bands of calorific value and limits of ash, water, and sulphur content, it could also outline a horizon of fuel development that could be used by coal companies to decide whether it would be worth investing in processing and cleaning facilities to capture a higher price for the same coal.

That this concept of fuel was no simple representation of a property in coal but a claim to coal's orderly behaviour in production can best be seen from the fall of sulphur: Once the principle of fire, the 'most active principle' that imbued all combustibles, and the substance that conferred activity on all bodies, it was demoted by fuel science to an 'accidental' and even detrimental element.[123] Unlike ash and water, however, pyrites and other forms of sulphur present in coal do *increase* the calorific value,

123 Ku-Ming Chang, 'Fermentation, Phlogiston and Matter Theory: Chemistry and Natural Philosophy in Georg Ernst Stahl's *Zymotechnia Fundamentalis*', *Early Science and Medicine* 7, no. 1 (2002): 48–9.

but they do so in a chaotic, unreliable way reminiscent of its association with the principle of fire: Apart from pollution, the most important reason for measuring sulphur content today is precisely its ability to 'spontaneously' generate heat, unwittingly ignite the coal material upon transport and storage, and destroy its use value. Fossil companies may have found a way to isolate and sell some of the sulphur as a by-product, and they may track coal's condition more closely than ever, but they still don't fully control its behaviour. Industrial capitalism searches in nature not for outbursts of activity but for a reliable, disciplined labour that nature by itself does not provide.

4

Electric Resistance

If we imagine the big problems solved, it is certain that electricity will become a consumer good on a par with bread, coal, iron and petroleum. But while these are hindered by all kinds of local and individual conditions in circulation and operation, electricity will achieve a universality such as gold has now, because a certain amount of electrical energy is the same here and the same in America and Australia. This, too, is a manifestation of the universal character of electricity. The simplicity that electricity will acquire as a commercial product makes it eminently suitable for large-scale operation, and it is certain that large-scale enterprises will one day emerge to take over the generation and distribution of electricity for significant areas, even entire countries. It will not be difficult for these companies to monopolize the trade in electricity for larger and smaller areas, and since it is not to be expected that the coexisting companies will fight each other, it is certain that the trade in electricity will one day come under the control of a few people. The only way to prevent this is to nationalize the production and distribution of electricity and thus also the power sources – and that is what will reliably happen.

– Arthur Wilke, *Die Elektrizität, ihre Erzeugung und ihre Anwendung in Industrie und Gewerbe*

In fin de siècle Berlin, electric light was still a luxury for the few. The many thousand wiremen, mechanics, and technicians employed by the city's booming electrotechnical industry could not hope to eat their supper under the light of a glowing wire. For some, however, cunning and technique made it possible to partake in the privilege as poorly adapted property rights only barely safeguarded electricity as a commodity. In a typical Berlin dwelling, where the bourgeoisie lived at the front and workers in the side or rear buildings, electrical wires connecting the rich to one of the capital's central stations were within the reach of the proletariat. In 1899, a Berlin mechanic was sued for having tapped an electrical wire to light his room. Yet the Imperial Court of Justice, the Reichsgericht in Berlin, had difficulty framing his deed as a theft, which was legally defined as the taking of an object without permission. As a disembodied service, the court ruled, electricity was not a thing taken away from someone, and so the accused was found not guilty. Still, contemporary observers agreed that a crime had occurred: How could this be different from tapping a gas line or stealing coal? The contentious verdict led to the adoption of a new law only a year later, in which tapping wires was criminalized as the 'diversion of electrical work'.[1] While this episode has itself become a trope to explain the elusive materiality of electricity, its significance for a history of the energy economy more broadly has not yet been noted: Working nature could now be owned and transferred; it had taken on an enforceable property form.[2]

Of course, the Prussian legislation confirmed only the enforceability of an existing property regime. Utilities had long charged their customers for the electricity used, measured by clocks and meters, regardless of its problematic status as a legal object. What power plant generators produce is a controlled drift of electrons of a certain speed and number within a

1 Gustav Siegel, *Die Elektrizitätsgesetzgebung der Kulturländer der Erde* (VDI-Verlag, 1930), 91.

2 Gretchen Bakke cites a version of it from Arkadii Markin's *Power Galore* (1961), where the episode takes place in imperial Russia; Gretchen Bakke, 'Electricity Is Not a Noun', in Simone Abram, Brit Ross Wintereik, and Thomas Yarrow, eds, *Electrifying Anthropology: Exploring Electrical Practices and Infrastructures* (Routledge, 2020), 25.

metallic wire. When an incandescent bulb was lit, the electrons passed through a filament – a finer wire of another material, often carbonized plant material – where they triggered small ripples in the atomic structure, producing heat and light. Electrons were neither created nor lost in the process. The electric field set them into a very slow, meandering motion, everywhere, instantaneously. Nothing tangible moved from the dynamo through the wire all the way to the electrical device. Yet the electric field constituted an invisible work potential that stood in a certain relation to the coal burned under the boiler of the steam engine that drove the dynamo and to the light and power into which it was ultimately transformed.

In nature, electricity manifests itself only in rare and elusive phenomena of charge and discharge. There is no natural electric current to which machines could cling; indeed, nature produces no stable electrical currents at all. 'Natural electricity,' wrote William Stanley Jevons when contemplating substitutions for coal in 1865, 'possesses all the characteristics of uncertainty and extreme irregularity, which are most opposed to utility.'[3] It becomes useful only in its socialized, tamed form: the electric current, an object that exists only in a technical system called circuit or grid. Electricity is comprehensively technically mediated – it is a 'technophenomenon' stabilized for experimentation and exploitation by natural scientists and entrepreneurs.

Precisely because it is unnatural can the current become nature's work in the purest, most reified form. The electric current does not embody its natural history; it does not reflect the raw material from which it was made, as products of coal and oil always do to some extent. At a time when there was still no reliable classification of global coals, its technical containment made electricity into a particularly universal good. 'Whereas [coal and petroleum] are limited in their circulation and application by all kinds of local and individual conditions,' predicted a German book on the commercial prospects of electricity in 1893, 'electricity will achieve a universal validation comparable to gold; as a certain unit of electricity is

3 William Stanley Jevons, *The Coal Question* (Macmillan & Co., 1866), 128.

the same here as in America and Australia.'[4] But electricity was not just any pure, raw material; it became the energy commodity par excellence. 'It is through the wholesale introduction of the electric current as a practical agent that the thing called "energy" has become a commercial commodity as it had before become a scientific measure', wrote John Merz, vice-chairman of the Newcastle-upon-Tyne Electric Supply Company and philosopher of the electric age, in 1903.[5] And the British economist Hugh Quigley called electricity 'an absolute in itself, incapable of modification or dilution.'[6] Consumers no longer bought a ton of coal and then had to deal with a widely fluctuating heating value; they could now purchase precisely the work per unit of time that they needed. Electricity can be scaled easily, divided into smaller amounts of work, and dovetailed to the task for which it is needed.

Crafting Currents

Forests can be enclosed, coal seams dug out, but electricity cannot so easily be captured. If there are no regular currents in nature, how did they come into being? Early experiences of natural electricity before its artificial production were rare and scattered. While charged amber and lightning were both commonly observed, the two were not recognized as related phenomena. When thermodynamics theorized machines that were already used in industry, the science of electricity and its instruments preceded commercial applications. Electricity was first created in the laboratory in order to be studied. The science of electricity is one of those fields where the technical became a mode of science, where scientists could no longer expect the instrument to leave the phenomenon

4 Arthur Wilke, *Die Elektrizität, ihre Erzeugung und ihre Anwendung in Industrie und Gewerbe* (Springer, 1893), 633.

5 John Theodore Merz, *A History of European Thought in the Nineteenth Century* (W. Blackwood and Sons, 1903), 156–7.

6 Hugh Quigley, 'Electricity as an Index of Industrial Production and Employment', in *The Transactions of the Second World Power Conference*, vol. 16, *World Problems of Power Economics* (VDI Verlag, 1930), 96.

untouched. Instruments began to do more than record and represent; they acquired themselves an epistemic function and reached into the very constitution of the phenomenon. The electric current is a '*phénomenotechnique*' – an object contained in technology. 'Without man on earth,' wrote Gaston Bachelard, 'there would be no electrical causality other than that which goes from lightning to thunder: light and noise. Only society can throw electricity into a wire.'[7] From the modern experimental scientists to the operators of power stations centuries later, electricity had to be handled as an electro*technical* fact, as an already theoretically invested entity.[8]

Electric utilities exploit, refine, and supply not a given resource but one they themselves create: The commodification of electricity hinges on the *technical production* of controllable currents. Electricity requires constant, uninterrupted form work in a manner coal, petroleum, and even gas do not. This technical containment means that theoretical relationships are already implicated in the realization of currents. The measurement of electrical work in a circuit is then a very different problem from the measurement of the calorific value in coal, a material that is not produced. Electricity is an already socialized nature, but it is not so thoroughly socialized that it would no longer reflect nature's resistance to human manipulation: As we will see, capitalists cannot produce and manipulate electricity as they wish. The design of their systems reflects the materiality of electricity, and profitable generation and management of currents continue to pose problems until today. While society has 'give[n] electrical phenomena the linear causality of a wire', concluded Bachelard, this only threw up the 'problems of branches' – a problem that would become critical in the exploitation of currents.[9]

With limited opportunities to observe and manipulate electrical phenomena *in natura*, the early science of electricity was experimental in the sense that its instruments did not so much emulate as produce its

7 Gaston Bachelard, *L'activité rationaliste de la physique contemporaine* (Les Presses Universitaires de France, 1965), 255.

8 Hans-Jörg Rheinberger, 'Gaston Bachelard and the Notion of "Phénomenotechnique"', *Perspectives on Science* 13, no. 3 (1 September 2005). 316.

9 Bachelard, *L'activité rationaliste de la physique contemporaine*, 255.

object of study. Luigi Galvani and Alessandro Volta used an electrostatic generator or an electrophore, apparatuses designed to separate positive and negative charges by friction. Only after electric charge was thus produced could current and its effects be studied. Electrical currents were realized in the laboratory long before their natural corollaries were discovered. But these crafted currents were short-lived and precarious laboratory phenomena, entirely dependent on the size and quality of batteries in which the manually separated charge was stored.[10]

Magnetic phenomena were better known but remained conceptually unrelated to electricity until Hans Christian Ørsted observed a magnetic needle's reaction to electricity in 1820. This interaction of magnetism and electricity prompted a further search for electromagnetic phenomena. A decade later, Michael Faraday described electromagnetic induction for the first time, the principle stating that moving electric charges create a magnetic field, and that a varying magnetic field, in turn, generates an electric field.[11] As experimental scientists, both Ørsted's and Faraday's discoveries were embedded in an experimental apparatus, enabling the creation of new phenomena. These new interactions, of stronger and more reliable currents, took place in and outside the laboratory. The earliest commercial application of Ørsted's and Faraday's electromagnetism was the telegraph, a device coding messages via punctuated interruptions of the electric circuit.[12] The magnetic force, triggered by an electric impulse sent through a coiled wire at the receiving end, yielded a recorded message.

10 Rudolf Stichweh, 'Technik, Naturwissenschaft und die Struktur wissenschaftlicher Gemeinschaften: Wissenschaftliche Instrumente und die Entwicklung der Elektrizitätslehre', in *Wissenschaft, Universität, Profession* (transcript, 2013), 89–90; J. L. Heilbron, *Elements of Early Modern Physics* (University of California Press, 1982); I. Bernard Cohen, *Franklin and Newton: An Inquiry into Speculative Newtonian Experimental Science and Franklin's Work in Electricity as an Example Thereof* (Harvard University Press, 1966), 286; Iwan Rhys Morus, *Frankenstein's Children: Electricity, Exhibition, and Experiment in Early-Nineteenth-Century London* (Princeton University Press, 1998).

11 P. M. Harman, *Energy, Force, and Matter: The Conceptual Development of Nineteenth-Century Physics* (Cambridge University Press, 1982), 30–3.

12 Ben Marsden and Crosbie Smith, *Engineering Empires: A Cultural History of Technology in Nineteenth-Century Britain* (Palgrave Macmillan, 2005), 178–225.

Soon, telegraph companies realized electric circuits at scales unseen in the laboratory. They were themselves experimental undertakings and provided invaluable data for scientific research, particularly in the domain of the use of materials and the velocity of signals.[13] Because knowledge of electrical measurements had a commercial value in telegraphy, particularly in submarine lines, engineers leaped ahead of scientists in the calculation and study of currents.

The first attempts at formalizing electrical phenomena took the science of heat as an inspiration, which from the late eighteenth century had been a highly mathematical and formal discipline. Electricity was conceptually associated with a potential. The realization of currents depended on the potential's difference along the wire and on the wire's conductive material. In 1827, Georg Ohm formulated a law in which the current flowing through a conductor was proportional to the voltage (the potential difference) applied to it (today $I = V/R$). He called the constant of proportionality 'resistance', denoting the degree to which a potential difference could realize a current in a given conductor. Materials differed in how they enabled or resisted the drift of electrons. Resistance varied by the type of material (today, we would relate it to the number of 'loose' electrons) and the length of the wire – it could be reduced by changing, shortening, or broadening the wire. Little appreciated at the time, Ohm's law became central to electrotechnical development over the course of the second half of the nineteenth century, as it expressed the insight that any realization of a controlled current was dependent on the resistance and potential difference over the entire network.

Among the avalanche of electrotechnical devices triggered by electromagnetism were also the first electromotors. For a short period in the 1830s, the supersession of steam appeared imminent. While most of the motors constructed at the time could propel only small objects, the most powerful, like Moritz von Jacobi's, carried a dozen passengers and a

13 Crosbie Smith and M. Norton Wise, *Energy and Empire: A Biographical Study of Lord Kelvin* (Cambridge University Press, 1989), 446–58; Bruce J. Hunt, 'The Ohm Is Where the Art Is: British Telegraph Engineers and the Development of Electrical Standards', *Osiris* 9, no. 1 (1 January 1994): 48–63.

200-kilogram zinc battery over the river Neva in 1838. James Joules, whose dedication to questions of efficiency compelled him to investigate electromagnetic engines as an economic source of power, showed that the electromagnetic engine was in principle more powerful than the steam engine. In practice, however, the low currents that could be gained from batteries, combined with the cost of the required zinc and battery fluids when compared to coal, made the engine impractical.[14]

Electromotors drew their power from a precarious source, a chemical reaction within a certain material, most often zinc. Yet the electromagnetic principle allowed for two possibilities: Not only would electricity excite a magnetic field, but moving electromagnets could conversely generate an electrical field. Electro-dynamos made use of this reversal, enabling the broader harnessing of motion for the purpose of creating currents. The 1860s saw the construction of electro-dynamos by Siemens, Gramme, Wheatstone, and other entrepreneurial engineers of the nascent electric power industry.[15] The dynamo could transform the mechanical power yielded from water or steam into electricity. Some, like Siemens's, were self-exciting, in the sense that they made double use of the electromagnetic effect: Moving magnets induced an electrical field, whose current ran through coiled wires around the magnets, thereby intensifying the magnetic and electrical fields. For all the improvements in battery design, only dynamos could generate a constant current powerful enough to galvanize material, illuminate streets and factories, and drive machinery – uses to which electricity was put without immediate commercial success.[16]

The crafting of the current is not just a historical precondition but one that underlies all later electrotechnical development. Currents appear and can be controlled only in technical systems where potential, current, and resistance relate to each other in a specific way. A change in one component

14 Crosbie Smith, *The Science of Energy: A Cultural History of Energy Physics in Victorian Britain* (University of Chicago Press, 1998), 60–1.

15 Wolfgang König, 'Elektrotechnik – Entstehung einer wissenschaftlichen Disziplin', *Berichte zur Wissenschaftsgeschichte* 10, no. 2 (1 January 1987): 83–93.

16 Wilfried Feldenkirchen, *Werner von Siemens: Inventor and International Entrepreneur* (Ohio State University Press, 1994), 87.

affects the others, and they are designed in a way that already reflects the theoretical conception of electric systems: Each part of an electric system, from the size of the wire, its material, the dynamos, and electric devices, has to be adapted to the system. This is also why a certain level of calculation and control devices were needed to realize a continuing electrical current in the first place: There is a quantitative relation between the capacity of the generator (the potential it generates) and the number of the arc-lights or capacity of an electric motor (the resistance) that must be considered before building the system. The matching of supply and demand – an economic principle in the oil and petroleum industries – is a technical requirement in electrical systems. This containment within a technical system is the basis of any commodification of the current; it enables the unique economies of central stations and the significance of electricity as a homogeneous, versatile energy commodity.

'A Fanciful Experiment'

To exploit the electrical current, it first had to be manufactured. Electrical manufacturing firms sold generation equipment – batteries, dynamos, water turbines, and generators – as well as electrical telegraphs, lamps, motors, and other devices that ran on electricity. Where flowing water could not be harnessed, the dynamo drew its motive power from a steam engine, making high-voltage electricity essentially a branch of the steam economy. Electrical manufacturers began to supply mines, factories, and public buildings with electric lighting systems of different kinds, and even tried to make electric lighting into a utility serving a variety of consumers – a 'fanciful experiment', as an op-ed in *Nature* put it in 1878.[17]

Electric manufacturing companies followed two marketing strategies. They sold the entire lighting system to industries planning to generate their own electricity and they operated electric utilities either alone or cooperatively. Industrial consumers usually owned the entire system, the

17 William Trant, 'The Divisibility of the Electric Light', *Nature* 19, no. 473 (November 1878): 52.

engines, generators, and wires, whereas utilities – often following the regulatory structure of gas lighting – held the rights to light a certain area for a certain duration. Some companies employed a kind of franchise system, forming the nested corporate structure which became typical of the electrical industry.[18] The manufacturing firm provided the utilities with equipment and the right to use it for a payment of rents and a share in the new company. A manufacturing company holding a stake in a utility often kept control over it until it had become profitable.[19] Utilities typically sold the arc-light service to a municipality at a fixed annual rate.[20]

Public street lighting dominates the memory of early electrification, even though isolated industrial stations likely made up the bulk of electrical production of the time.[21] Electric lighting systems developed around the arc-light – essentially a glowing break in a circuit. At the point of interruption, the electric current sprang from one small piece of carbon to another in the shape of an arc. What generated the light was not, as its inventor Humphry Davy thought, the arc as such, but the two pieces of carbon, smouldering in the atmosphere.[22] Just like gas lighting, the electric lighting industry was supplied with by-products of coal and petroleum. While lighting gas was derived from a specific type of bituminous coal, lighting carbon was generally from petroleum coke. From a technical perspective, these lamps were resistances within the electrical circuit, and the generator had to be chosen according to the type and quantity of lights to be supplied.

Apart from its aura of modernity, skilfully cultivated by the companies, arc-lighting held few advantages over gas. The first systems did not even allow for a series connection of several lights. The Siemens lamp, for example, required one generator for each lamp, making the system

18 Harold C. Passer, *The Electrical Manufacturers, 1875–1900: A Study in Competition, Entrepreneurship, Technical Change, and Economic Growth* (Harvard University Press, 1953), 19.

19 In this it anticipated later holding companies; ibid., 14–40.

20 Ibid., 82.

21 Wolfgang Schivelbusch, *Disenchanted Night: The Industrialization of Light in the Nineteenth Century* (University of California Press, 1988), 120.

22 Ibid., 52.

prohibitively expensive.[23] Manufacturing companies such as Grimme, Brush, and Siemens drove the development of dynamos.[24] Because early generators could not be regulated – that is, adjusted to variations in load – a second set of arc-lights (or any other resistance) had to be kept at the station to be connected whenever the street lighting was shut down.[25] Even in the best systems, workers had to replace the pieces of carbon nearly every two hours, a constraint that contributed significantly to the system's labour costs.[26] The electric systems at that time operated with a constant current, the voltage of the lamps was known, and all lamps were usually controlled by one switch. There was thus no intended diversity in load, voltage, or current. Indeed, in the late 1870s, some still believed the electric current to be indivisible.[27]

The only advantage arc-light systems had was in terms of the amount of light they could produce, which made it suitable for industrial and public spaces. Yet this blinding brightness, a hundred times lighter than gas lights, also made them unsuitable for apartments and most other interior spaces. The arc-light was thus limited to outdoor and very large spaces, where all lamps could be used to the same end, as in the illumination of streets, theatres, public buildings, or construction sites.[28] The disadvantage of gas – that it consumed oxygen – became a limit only when the electric light had normalized an abundance of brightness. Gas remained more flexible; it was cheaper and hard to displace, particularly in Europe where it was not rare for cities to operate their own gasworks.[29]

23 P. Strange, 'Early Electricity Supply in Britain: Chesterfield and Godalming', *Proceedings of the Institution of Electrical Engineers* 126, no. 9 (1979): 864.

24 James E. Brittain, 'The International Diffusion of Electrical Power Technology, 1870–1920', *Journal of Economic History* 34, no. 1 (March 1974): 108–21.

25 Passer, *The Electrical Manufacturers*, 63.

26 Ibid., 17.

27 Ibid., 82.

28 Schivelbusch, *Disenchanted Night*, 54–6.

29 Strange, 'Early Electricity Supply in Britain', 863; Leslie Hannah, *Electricity Before Nationalisation: A Study of the Development of the Electricity Supply Industry in Britain to 1948* (Macmillan, 1979), 7; Daniel R. Shiman, 'Explaining the Collapse of the British Electrical Supply Industry in the 1880s: Gas Versus Electric Lighting Prices', *Business and Economic History* 22, no. 1 (1993): 318–27.

Electrical manufacturers at the time were engaged in a high-risk gamble that creating new markets for their equipment would be profitable.[30]

Due to their fixed technical specification, these lighting systems were static and could not easily be expanded and adapted. Electrical engineers designed and planned the illumination of a specific street, building, or factory. They first calculated the number of lamps and circuits, and then the engines and generators needed to light them. The power of the machine generating this electricity was incidental to the light supplied, and it could be determined only by experiment and experience.[31] In burning, arc-lights consumed the two carbon components, and therefore distance between these poles grew. For that reason, arc-lights burned regularly only in circuits where everything else was held constant, making it difficult to expand the system.[32] Because arc-lights require a specific amount of voltage difference to induce a spark, single lights were extremely bright, and the number of lights in one circuit was limited by the voltage of the system. Expressed in commercial terms, the light sold could not be subdivided retroactively within a given system.

The further expansion of the electric lighting business rested on the 'question of subdivision', as it was called at the time. Unless electrical light could be sold in smaller packages, in which it could illuminate all rooms and corners of a building separately, it remained limited to the niche market of spectacle and celebration. Only by extending the electric station's supply to include private consumers, as William Siemens concluded after his water-powered arc-light station in Godalming shut down, were they 'ever likely to make a profit'.[33] In the 1870s, several lighting systems dealt with these problems by way of sophisticated lamp designs, each of which made it possible to connect lights in a series or in parallel. Because a glowing carbon filament did not require the high voltage difference of a spark, the

30 Passer, *The Electrical Manufacturers*, 28; John L. Neufeld, *Selling Power: Economics, Policy, and Electric Utilities Before 1940: Markets and Governments in Economic History* (University of Chicago Press, 2016), 22.

31 Friedrich von Hefner-Alteneck, 'Ueber die Theilung des elektrischen Lichtes', *Elektrotechnische Zeitschrift* 1, no. 3 (1880): 81.

32 Ibid., 84.

33 Cited in Strange, 'Early Electricity Supply in Britain', 866.

incandescent light was considered 'subdivision in perfection'.[34] By the end of the 1880s, the largest arc-light systems could electrify only around fifty lamps, while incandescent systems were able to cover thousands.[35]

One of those systems that were intended to overcome the limits of early lighting systems was Thomas Edison's. Through careful study of the gas industry, he had learned that most profit derived from sales to private consumers, and therefore electric lighting systems would have to reach them.[36] In 1876, Edison summoned engineers and mathematicians to Menlo Park to work on a new kind of system, modelled on gasworks, and consisting of a high-resistance incandescent light connected in parallel to a low constant-voltage dynamo. None of the major elements of Edison's system were new, but it integrated them into one workable platform requiring a series of novel features: protection against short circuits, sockets, switches, meters, a constant-voltage dynamo, and efficient low-voltage transmission. Instead of a system designed around the voltage that could sustain the number of lights required, Edison's system was designed around the current: In varying it, Edison was able to connect a large, flexible number of lights in parallel.[37] In Edison's incandescent lighting system, the components affected and depended on each other in a new way that still rested on but went beyond the balancing of voltage, current, and resistance. With each connected user and their pattern of utilization affecting the economic state of the whole, lighting systems had become an *economic* system.

This was the beginning of the electric utilities business – and Edison was not alone in it. The lighting systems of Edison, Westinghouse, and Thomson-Houston served, rather than one large-scale public consumer at a time, a public composed of individual consumers. Edison presented his

34 Hefner-Alteneck, 'Ueber die Theilung des elektrischen Lichtes', 91.

35 T. C. Martin, 'The Electric Light Industry in America in 1887', *Electrician* 18 (1887): 315.

36 Passer, *The Electrical Manufacturers*, 82; Harold C. Passer, 'The Electric Light and the Gas Light: Innovation and Continuity in Economic History', *Explorations in Entrepreneurial History* 1, no. 3 (1949): 2–3.

37 Passer, *The Electrical Manufacturers*, 80–1; Thomas P. Hughes, *Networks of Power: Electrification in Western Society, 1880 1930* (Johns Hopkins University Press, 1993), 30–2.

system at the World Exhibition in Paris in 1881, where it was enthusiastically received. The newly founded Compagnie Continentale d'Edison marketed his patents in Europe and spawned many new companies exploiting the patents in a specific country, such as Emil Rathenau's Allgemeine Elektrizitätsgesellschaft (AEG, then Deutsche Edison). However, the central stations operating in the 1880s were of various designs. Ensuing patent and supply relations did not prescribe their precise layout, even though they fostered international standardization and exchange. Some stations ran on alternating, others on direct current; they differed in the size and type of boilers and generators, in their organization of system control, in their consumers (usually a combination of arc-lamps and incandescent lamps), and in the way they charged them (by meter or contract).

For all the commercial problems of electrical lighting and its improvement, not everybody was convinced the future of the electric manufacturers lay in lighting utilities.[38] Central stations remained economic experiments and risky undertakings, especially as long as they offered a service that already existed. Sharing the problem of other new, capital-intensive undertakings, they had to account for and earn fixed capital costs, without many experiences of the system's typical wear and tear. It was not rare for the early stations to operate for only a couple of years. Two British stations, one mixing arc and incandescent lights, built by Siemens in 1881, and one by Edison in 1882, closed after two years due to insufficient demand.[39] The Pearl Station burned down after less than a decade but was rebuilt the following year because of its significance 'as showpiece and inducement for others to purchase Edison franchises and Edison equipment'.[40] Whether, and how, central stations could be profitably managed was still an open question.

Managers and electric engineers went to great length to prove that electric lighting was as cheap, or cheaper, than gas lighting. While

38 Werner von Siemens did at the time not believe in the future of electric lighting at all; Passer, *The Electrical Manufacturers*, 79–81; Jürgen Kocka, 'Siemens und der aufhaltsame Aufstieg der AEG', *Tradition: Zeitschrift für Firmengeschichte und Unternehmerbiographie* 17, no. 3/4 (1972): 128.

39 Hughes, *Networks of Power*, 55–62.

40 Ibid., 43–5.

American companies could venture into a direct competition with gas-lighting companies, European companies found themselves in a more difficult situation. As a larger share of European municipalities had an interest in or operated their own gasworks, they were generally more hesitant to allow competition. At the very least, if an electric lighting system was to replace gasworks, it had to be able to yield a comparable profit for the city.[41] Discussions and comparisons of the cost of central station service were prevalent in electrotechnical journals over the late 1880s and early 1890s.[42] When it came to production costs, electric light was not necessarily more expensive than gas light. Indeed, these numbers could even be tweaked in favour of electric light when they were related to the cost of the intensity of light – a relation foregrounded by the electrotechnical industry, but of questionable practical value (there are few purposes that benefit from as much light as possible). Those cost calculations often conveniently excluded capital costs, the money needed for interests, repairs, and amortization. Capital costs amounted to a large share of the total cost of electric service, a share that decreased with the size of the station.[43]

---

41 Daniel Shiman puts the number of municipal gasworks before electrification at 30 per cent in Great Britain, against almost total private supply in the US. In Germany, where some concessions stipulated that gasworks would become public property after a fixed period, states or municipalities owned and operated 50 per cent of the gasworks and held a share in many more. Shiman, 'Explaining the Collapse of the British Electrical Supply Industry', 321; Hans Geitmann, *Die wirtschaftliche Bedeutung der deutschen Gaswerke* (R. Oldenbourg, 1910), 8–9.

42 See, for instance, George Forbes, 'Some Electric Lighting Central Stations in Europe, and Their Lessons', *Journal of the Institution of Electrical Engineers* 18, no. 78 (1889): 161–97; W. Fritsche, 'Sur les stations centrales d'éclairage électrique', *La lumière électrique* 10, no. 19 (1888): 251–63; Gisbert Kapp et al., 'Discussion on Professor G. Forbes's Paper on "Some Electric Lighting Central Stations in Europe, and Their Lessons"', *Journal of the Institution of Electrical Engineers* 18, no. 79 (1889): 211–40; Max Meyer, 'Die Entwickelung der städtischen Elektrizitätswerke', *Elektrotechnische Zeitschrift* 16, no. 2 (1895): 26; R. Rühlmann, 'Einige Gesichtspunkte, welche bei der Errichtung von Elektrizitätswerken in Betracht zu ziehen sind', *Elektrotechnische Zeitschrift* 9, no. 13 (1888): 309–22.

43 Fritsche, 'Sur les stations centrales d'éclairage électrique', 259.

While central stations had successfully subdivided electric light – a precondition for further commodification – they ran into economic problems of their own. Both gas-light and arc-light systems shared the problem of high capital costs. However, the latter was tailored to light a building or street for a certain length in time, stipulated in the contract, and the former could at least react to consumers' demand by sending less gas through the pipes and storing it for later days. Central stations, in contrast, could neither count on a certain demand nor adapt their production to consumption (they could do so only in a stepwise manner, by bringing entire generators online). All the while interest on capital had to be paid regardless. The economic theorizing of central stations evolved around this problem.[44]

Which station design and mode of operation best accommodated those problems remained contested into the 1890s. It seems that most central station managers tended to reduce costs rather than create demand: As late as 1894, a German electrotechnical engineer argued that limiting operational costs was more economically rational than any effort at achieving full utilization of capital employed and designing the

44 Throughout the 1880s and 1890s, central station managers understood themselves to be mainly in the business of illumination. However, due to the surplus steam and electricity they almost necessarily produced, they can also be seen as pioneering the energy service company. Edison's Pearl Street Station in Manhattan sold not only light but also heat to households and processed steam for small urban industries, making it an early site of cogeneration. Robynn Andracsek and Minda Nelson, 'Bringing Power Back to Town', *Power Engineering* 114, no. 11 (1 November 2010): 12–13; Joseph J. Cunningham, 'Forgotten Pioneer: Leo Daft and the Excelsior Power Company [History]', *IEEE Power and Energy Magazine* 16, no. 4 (2018): 108–20; Pearl Street was a not uncommon variant of station design, just as was the case with isolated stations selling their excess steam and electricity to utilities, or coal companies venturing into the sale of steam; S. M. Bushnell, 'Central Station Operation of Steam Plants in Connection with Lighting Company's Service', *National Electric Light Association, 32nd Convention*, vol. 2, 1909, 778–815; Hannah, *Electricity Before Nationalisation*, 181. Selling steam could also be a way to bypass the regulations of the steam economy. When the Rheinisch-Westfälische Elektrizitätswerk (RWE) was founded in 1898, Hugo Stinnes, owner of the brown coal mine supplying the RWE and one of the most powerful people on its board, decided to sell steam, not coal, in order to avoid being charged a share by the Rhenish-Westphalian Coal Syndicate.

most cost-efficient average lamp-hour.[45] To do so, the station's size should be carefully planned. Edison himself had also envisioned a more conservative system of block stations with small and flexible machinery that could adapt to fluctuations of the load.[46] Over the 1890s, however, attention converged on a higher utilization of capital-intensive machinery and return on investment.[47] A German engineer approvingly quotes the new motto of the industry that he picked up on a business trip to the US: 'Between the heater's coal shovel and the dynamo's pole binders is the place where money is made or lost in the central station business.'[48] This, of course, was exactly where fixed capital was concentrated.

Drifts of Electrons, Streams of Revenue

Even though the electrical current could now be called into existence and divided at will, its ability to support a stream of revenue was still precarious. Central stations kept failing – especially during the financial crisis of the 1890s. Unlike with coal or petroleum markets, electrical systems integrated production, transmission, and consumption in a way that it had to be balanced out at the timescale of milliseconds. Within this highly technical system, built on the mastery of the current, appeared a new problem of control: In fact, the flow of electricity was not unlike the autonomy of a stream of water, whose variation could be studied but was out of the capitalist's reach. Precisely because central station operators could not control demand directly (as in industrial stations) or fix it by way of a contract (as in earlier lighting systems), they had to find a

45 A. Müllendorf, 'Ueber die Beurtheilung der Rentabilität elektrischer Anlagen', *Polytechnisches Journal* 292 (1894): 110.

46 See, for Samuel Insull's memories of a conversation with Emil Rathenau about station size, Samuel Insull and William Eugene Keily, *Central-Station Electric Service: Its Commercial Development and Economic Significance as Set Forth in the Public Addresses (1897–1914) of Samuel Insull* (Priv. Print., 1915), 136–7.

47 Norbert Gilson, *Konzepte von Elektrizitätsversorgung und Elektrizitätswirtschaft: Die Entstehung eines neuen Fachgebietes der Technikwissenschaften zwischen 1880 und 1945* (GNT-Verlag, 1994), 35–6.

48 Rühlmann, 'Einige Gesichtspunkte', 312.

way to shape and develop it indirectly. Only the second generation of electrical entrepreneurs, men like Samuel Insull, Oskar von Miller, and Emil Rathenau, seized the economic opportunity that lay in electricity's technical containment and punctuated temporality.[49] By focusing on full utilization over time, engineers and managers transformed the economic restriction posed by electricity's specific materiality into an opportunity to yield a smoother stream of revenue.[50]

By the turn of the century, central stations built by a handful of mostly American and German electric companies operated in cities around the globe and penetrated ever more areas of social life. Apart from providing light to the wealthiest neighbourhoods, they had slowly moved into the traction and motor business. With alternating current and high-voltage transmission, systems could grow from hundreds of metres to several kilometres at the turn of the century. A manufacturing company, or its branch focusing on lighting systems, could no longer realize such projects of this scale and complexity alone. What is more, it would rather not take on the debt and risk that came with it. This is why manufacturers who had ventured into the central station business began to create an intermediary organization between the seller and buyer of a utility in close contact with financial backers.[51] These intermediaries, set up for the sole purpose of creating and financing electrical utilities, appeared in different forms, ranging from the *Unternehmergeschäft* in German-speaking countries, to the typical American holding companies.[52] Edison cooperated with J. P.

49 According to the memories of Oskar von Miller, then president of the German Edison Company, an awestruck Edison watched the Berliner Elektrizitätsbetriebe's large machinery in 1891, which made '8 Mark 30 Pfennig for every rev!'; Oskar von Miller, 'Die Entwicklung der Elektrotechnik', *Gaea: Natur und Leben* 42, no. 10 (October 1906): 616.

50 Bakke, 'Electricity Is Not a Noun', 33.

51 Kocka, 'Siemens und der aufhaltsame Aufstieg der AEG', 129–30; Thomas P. Hughes, 'The Electrification of America: The System Builders', *Technology and Culture* 20, no. 1 (1 January 1979): 153–4.

52 Bernhard Stier, *Staat und Strom: Die politische Steuerung des Elektrizitätssystems in Deutschland 1890–1950* (Verlag Regionalkultur, 1999), 44. Holding companies have forerunners in the financing of gasworks and railways; Serge Paquier, 'Swiss Holding Companies from the Mid-Nineteenth Century to the Early 1930s: The Forerunners and Subsequent Waves of Creations', *Financial History Review* 8, no. 2 (October 2001): 163–82.

Morgan, while Siemens and AEG joined forces with Swiss banks to establish Elektrobank and Indelec respectively.[53] The activities of these companies went far beyond the coordination of financiers. To make sure that money for money would be forthcoming, they planned, projected, and supervised undertakings to ensure their profitability. Holding companies, engineering firms, and manufacturing companies aligned their streams of electrons and revenues by formalizing the economic relations of electrical systems.

The 'Long Depression' in the late nineteenth century did not spare the electrotechnical industry and central station business. The 1890s, in particular, saw the death of many smaller companies and increased corporate concentration. In the US, the panic of 1893 – a broader economic crisis – threw the industry in disarray and caused a number of central station bankruptcies, reinforcing the financial industry's grip on the business and triggering a buy-up of flailing companies.[54] The period between 1897 and 1902 is known as *Elektrokrise* in Germany: A period marked by overproduction and a satiation of lighting and traction markets, increased raw material prices (coal prices surged by 50 per cent in 1899), and the flight of capital. 'Capital is no longer rushing to invest in electrical assets as it did a few years ago', observed the Elektrobank's annual report for the year 1900/1 dryly.[55] The entire period of fierce competition had repercussions throughout the electrotechnical industry.[56]

The crisis of overproduction was primarily one of an abundance of electrical machinery, not one of electricity. Still, the formalization of central station cost accounting from the 1890s onwards must be seen in

53 The American system of holding companies differed insofar as the holding company kept controlling shares in the operating companies, at least before the Federal Trade Commission's intervention; Neufeld, *Selling Power*.

54 Jeremiah D. Lambert, *The Power Brokers: The Struggle to Shape and Control the Electric Power Industry* (MIT Press, 2015), 10; Patrick McGuire, Mark Granovetter, and Michael Schwartz, 'Thomas Edison and the Social Construction of the Early Electricity Industry in America', in Richard Swedberg, ed., *Explorations in Economic Sociology* (Russell Sage Foundation, 1993), 233.

55 Cited in Felix Pinner, *Emil Rathenau und das elektrische Zeitalter* (Akademische Verlagsgesellschaft, 1918), 228.

56 Ibid, see, for repercussions in Great Britain, J. F. Wilson, *Ferranti and the British Electrical Industry, 1864–1930* (Manchester University Press, 1988), 78–9.

the context of the industry's crisis and its dependence on public capital.[57] This is most explicit in the Elektrobank's annual report from the year 1901/2. Given that the most viable projects in the field of central stations and tram service had already been done, the report suggested that

> further activity in this direction will therefore either have to extend to more distant, politically and economically less developed countries or, by reducing the cost of installation and operation, seek to make the advantages of electric lighting and traction accessible to those communities which have not hitherto been considered sufficiently worthwhile for such facilities.[58]

With every large European city electrified, every horse-tram replaced, the electric business could only intensify its overseas business (subject to all sorts of risks) or find a way to make central stations more profitable.

Load management became the theory that described how central stations could be operated and developed profitably.[59] Over the following two decades, manuals on central station service appeared that all highlighted the role of load management and rate making.[60] A theoretical reflection on the cost of central station service had already begun in the 1880s – and it is perhaps no accident that the discussion was most sophisticated in the UK, where competition with gas lighting was the fiercest. John Hopkinson and J. A. Fleming, two British electrotechnical

57 Luciano Segreto, 'Financing the Electric Industry Worldwide: Strategy and Structure of the Swiss Electric Holding Companies, 1895–1945', *Business and Economic History* 23, no. 1 (1994): 164; Hughes, 'The Electrification of America', 154–5.

58 Cited in Pinner, *Emil Rathenau und das elektrische Zeitalter*, 233.

59 Stier, *Staat und Strom*, 51.

60 Insull and Keily, *Central-Station Electric Service*; Georg Klingenberg, *Bau großer Elektrizitätskraftwerke*, vol. 2, *Verteilung elektrischer Arbeit über große Gebiete* (Julius Springer, 1914); Georg Klingenberg, *Bau großer Elektrizitätskraftwerke*, vol. 1, *Richtlinien für den Bau großer Elektrizitätswerke* (Julius Springer, 1913); Georg Klingenberg, *Bau großer Elektrizitätskraftwerke*, vol. 3, *Das Kraftwerk Golpa* (Julius Springer, 1920); C. H. Merz and Wm. McLellan, 'Power Station Design', *Journal of the Institution of Electrical Engineers* 33, no. 167 (July 1904): 696–742; H. G. Solomon, *Electricity Meter Practice* (Charles Griffin & Company, 1923).

engineers, conducted one of the earliest formal cost analyses of electric service at the London Holborn Viaduct, Edison's second station, in 1882, which kept running up losses and eventually had to shut down. In Germany, the dominant formula to calculate profitability of a system consisted of relating total system costs to lamp-hours used – a ratio that reflected capital utilization but did not account for the difference in fixed and operational costs, even though each additional lamp-hour affected system costs very differently, depending on the timing.[61]

The load factor probably first appeared in a paper by Rookes E. B. Crompton on the cost of central station service in 1891 and was quickly picked up by others in the industry.[62] The term 'load' originated in the study of engine efficiency earlier in the nineteenth century. In the steam economy, 'duty' was a standard measure of engine performance. As Crosbie Smith argues, James Joule had distinguished between

> 'duty' as 'pounds raised per second of time to one foot in height' and 'economical duty' as 'pounds raised to the height of one foot by the agency of one pound' of coal in the steam engine or of one pound of zinc consumed in the battery used to power the electromagnetic engine.[63]

This latter meaning migrated into electrical engineering by way of comparison between electromagnetic and steam engines. Whereas

61 Müllendorf, 'Ueber die Beurtheilung der Rentabilität elektrischer Anlagen'; F. Ross, 'Ueber die Entwickelung elektrischer Centralstationen', *Elektrotechnische Zeitschrift* 13, no. 19 (1892): 255. The absence of the watt in this formula also reflected the fact that systems were not yet seen as selling electrical work, even though they were technically designed around it.

62 R. E. B. Crompton, 'The Cost of the Generation and Distribution of Electrical Energy', *Minutes of the Proceedings of the Institution of Civil Engineers* 106, no. 1891 (January 1891): 2–32; John Hopkinson, 'On the Cost of Electric Supply (1892)', in Edison Illuminating Company, ed., *The Development of Scientific Rates for Electricity Supply; Being Reprints of Selected Original Rate Papers* (Edison Illuminating Company of Detroit, 1915), 8–10; G. P. Watkins, 'A Third Factor in the Variation of Productivity: The Load Factor', *American Economic Review* 5, no. 4 (1915): 753; C. P. Feldmann, 'Ueber die Faktoren, welche die Rentabilität der Elektricitätswerke beeinflussen, Teil 1', *Elektrotechnische Zeitschrift* 18, no. 51 (23 December 1897): 779–81.

63 Smith, *The Science of Energy*, 57.

Joules varied the source of power and kept load constant to measure economic efficiency, electrical engineers thought of the central stations as a constant source of power and varied load. In doing so, they essentially treated the electrical system as a single machine at a certain capacity reaching its highest efficiency when load matched capacity: The load factor denotes the (average) power consumed in relation to maximum capacity over a certain stretch of time, or the degree to which fixed capital is utilized at a certain point in time. The load factor is a straightforward measure of capital utilization in industries in which capacity can be used as a proxy for capital outlay, and it gained a critical importance in the electric industry as electricity could not be stored.[64] By the late nineteenth century, with the increasing pressure to bring down central station costs, the load factor had become the most important operating principle of central stations, an indicator to improve profitability of electrical power plants and to maximize return on investment.[65]

The load factor not only described but also made it possible to envision future states of the system. The diagram was the medium in which those states could be represented: It showed that the systems reached their peak – the maximum of load for which the machines had been designed – only for a very short period, usually on an evening in winter, when the need for light was the highest. A paper that summarized the international discussion on central station viability in 1888 found that many calculations came down to an average load factor of 15 per cent, which remained typical for lighting-load-dominated central stations.[66] It was not uncommon for stations to reach peak demand for only thirty minutes a day.[67] The capacity to supply this maximum demand sat idle for the rest of the time – expressed in the volume of the

64 Similar measures had before been applied in other capital-intensive infrastructure enterprises; Hughes, *Networks of Power*, 219.

65 Hughes, 'The Electrification of America', 150–1.

66 Rühlmann, 'Einige Gesichtspunkte'; Stier, *Staat und Strom*, 52.

67 Otto M. Rau, 'The Changed Significance of Load Factor', in American Academy of Political and Social Sciences, ed., *Giant Power: Large Scale Electrical Development as a Social Factor*, Annals of the American Academy of Political and Social Sciences, vol. 118 (1925), 133.

area above the curve – while interest on the capital expenditure had still to be paid. This is why 'reasonable managers have realized the great economic benefits that can be achieved by reshaping this curve', as AEG's Georg Klingenberg put it nonchalantly in 1913.[68] The load curve could be changed in several ways: First, by changing the time and composition of consumption in a given system through carefully crafted tariffs, new uses for electricity, or the integration of new groups of users. Second, by interconnecting existing electric systems, which could lead to an overall improved resource use and a lowering of reserve capacity and was practised since the 1890s.

Load diagrams could be charted for the system as a whole or – with the right meters – for individual consumers. While earlier meters had measured the length of time over which a lamp burned (the lamp-hours), load management depended on new metering and information-gathering devices.[69] Rather than time, they now measured the power used (watt-hours), the peak in a customer's demand, or the length of time over which the maximum of devices were used simultaneously. By doing so, operators transformed individual behaviour into a comparable pattern that bore an economic relationship to the overall profitability of the station.[70] In the medium of the load diagram, the social and temporal variation of consumption that is relevant for electrical systems takes form. Winter and summer loads appeared distinct in the load diagrams, as did the 'household load', 'industrial load', 'traction load', each of which constituted a recognizable pattern across many different stations, cities, and countries.

On this basis, tariffs could be designed in a way to incentivize consumer behaviour that would improve the system's utilization. While prices had first been oriented towards undercutting gas lighting, managers soon realized that they had to recoup capital outlays to avoid closures. Debates over the rate of depreciation, the unit of charge (lamp-hours or kilowatt-hours), the ratio of fixed to operational costs, and the

68 Klingenberg, *Bau großer Elektrizitätskraftwerke*, vol. 3, 1.

69 Hughes, 'The Electrification of America', 150.

70 Feldmann, 'Ueber die Faktoren, welche die Rentabilität der Elektricitätswerke beeinflussen, Teil 2',

construction of a rate system reflecting these relations grew in intensity over the 1890s.[71] Rates not only differed with regard to a customer's share of capital costs but were also used to gain new loads and to expand the system. Around the turn of the century, central stations encouraged the use of electromotors and sought to supply small industries to balance the lighting load, concentrated as it was in the morning and evening hours.[72] To do so, prices had to undercut competitive technologies, such as individual steam power, gas motors, and pressured air systems. To gain urban industries as consumers, operators either reduced power tariffs directly or based them on the customer's load factor or the percentage of time over which the maximum demand for service could be extended. The latter had the advantage of generally encouraging a more constant demand. As systems developed, the design of tariffs never came to an end.

The inclusion of motor load was only one way to diversify electricity use away from the station's peak. Diversification referred to the inclusion of consumers with different load profiles and peak loads into a single system, as well as their 'education', in the words of the industry, to shift their peak. The targeting of industrial load was an attempt to diversify, but the strategy was extended further to railroads, streetcars, and other businesses. Every trade had its own electric 'loadprint', defined by the shape of the curve, the average and peak loads. Few went as far as Samuel Insull, director of Commonwealth Edison Company, in formalizing this relation. In his aggressively expanding Chicago utility, Insull introduced a concept called the 'diversity factor' to indicate the degree of diversion each customer exerted on the system's load.[73] To improve this factor one needed to obtain 'customers who make the maximum demand on you for your product at different hours of the day, or different days of the

71 Valery Yakubovich, Mark Granovetter, and Patrick McGuire, 'Electric Charges: The Social Construction of Rate Systems', *Theory and Society* 34, nos 5–6 (2005): 579–612.

72 Franz Ferrari, 'Die Entwicklung der Elektrizitätszähler und ihre Verknüpfung mit der Elektrizitätswirtschaft', *Technikgeschichte: Beiträge zur Geschichte der Technik und Industrie* 28 (1939): 72.

73 Insull and Keily, *Central-Station Electric Service*, 80–1.

week, or different weeks of the month, or different months of the year'.[74] The degree to which each customer differed, of course, depended on all other customers.

Another way to rationalize systems was through their interconnection, a diversification of consumption *and production*. It first referred to the interconnection of smaller systems, which could together reach a higher capacity and better load, or the integration of isolated plants into the utility grid.[75] Under ambitious managers, however, electric utilities could nearly monopolize energy production and consumption within a city or region (see Fig. 4.1). In Insull's understanding, 'the proper function' of all electric utilities was 'the turning of all the wheels of industry' and 'the operation of all the transportation systems in the city' – and beyond. Utilities should strive to be 'the sole producer of electrical energy in a given community'.[76] Not all managers pursued the strategy of expansion as consequently as Insull, who, after a Federal Trade Commission investigation into his pyramid scheme, had failed badly by the 1930s.[77] As a rule, company-led expansion into interconnected regional systems was constrained by profitability and government regulation (especially eminent domain, anti-trust, and utility regulations).

During and after World War I, states grew more directly involved in the electrical power industry and established centralized institutions to oversee production and distribution. Industry was often coerced into giving up isolated stations and joining central stations; grids were gradually connected to each other, and new institutions and commissions for the allocation of electricity were formed.[78] Thrown back on a limited

74 Ibid., 128–32. The indicator was used beyond the US; see Feldmann, 'Ueber die Faktoren', 790.

75 Julie A. Cohn, *The Grid: Biography of an American Technology*, reprint ed. (MIT Press, 2017), 13–26.

76 Insull and Keily, *Central-Station Electric Service*, 142.

77 Lambert, *The Power Brokers*, 38–49.

78 Jonathan Coopersmith, 'When Worlds Collide: Government and Electrification, 1892–1939', *Business and Economic History On-Line* 1 (2004): 12–13; William J. Hausman, Peter Hertner, and Mira Wilkins, *Global Electrification: Multinational Enterprise and International Finance in the History of Light and Power, 1878–2007* (Cambridge University Press, 2008), 127–8; Hannah, *Electricity Before Nationalisation*.

national territory, 'domestic' fuels gained importance, even if they were of a considerably lower grade, such as peat and lignite.[79] Some states even nationalized the electric power industry.[80] In order to rationalize consumption during the war, national surveys of electrical use were made; they were often the first such studies of consumption beyond those of isolated central stations.[81] During the war and into the interwar years, many governments set up national electricity commissions in order to secure and ration supply.

Engineers and politicians determined that the nation, rather than region or city, was to become the basic unit for postwar power development.[82] At this level, interconnection entailed the integration of entire power plants under regional or national systems. The technology of long-distance high-voltage transmission revived hydraulic power, making it possible to combine hydroelectricity and fuel-powered electricity in one large power pool.[83] The arguments for interconnection essentially followed the principle of 'diversification' established in central station management. By connecting different areas and consumers with different load profiles on a national grid, a better utilization and broader supply could be achieved.[84] At the time, it was widely accepted that the electrical supply industry was a 'natural monopoly' and that interconnection and centralization should be encouraged.[85]

79 Hausman, Hertner, and Wilkins, *Global Electrification*, 128.

80 Ibid., 253–6.

81 Ibid., 127.

82 Ibid., 129.

83 Hughes, *Networks of Power*, 264, 324–62; Astrid Kander, Paolo Malanima, and Paul Warde, *Power to the People: Energy in Europe over the Last Five Centuries* (Princeton University Press, 2014), 266; Vincent Lagendijk, *Electrifying Europe: The Power of Europe in the Construction of Electricity Networks* (Amsterdam University Press, 2008), 26–8, 40.

84 Lagendijk, *Electrifying Europe*, 22, 47; Hughes, *Networks of Power*, 462.

85 Bernhard Stier argues that the concept of a 'natural monopoly' resting on the technical-economic nature of electricity made state intervention, in the form of regulation or direct involvement, acceptable to liberal elites; Stier, *Staat und Strom*, 60–7.

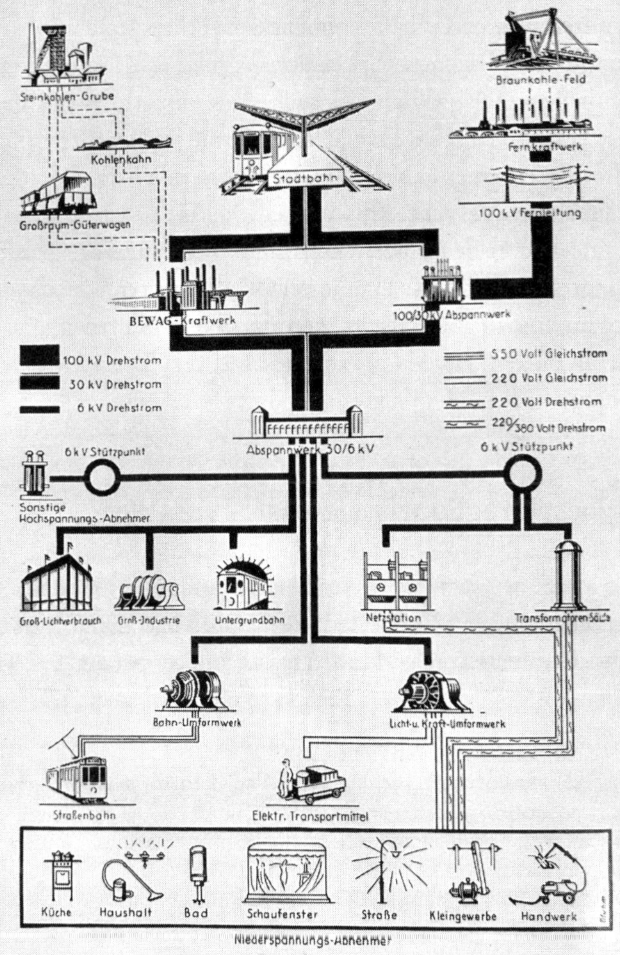

Bild 65. Schema der Stromversorgung Berlins

Figure 4.1. Berlin's electrical supply system, integrating lignite- and coal-fired power plants as well as a diversity of urban consumers (c. 1930). Conrad Matschoss, E. Schulz, and Arnold Theodor Gross, *50 Jahre Berliner Elektrizitäts Werke, 1884–1934* (VDI-Verlag, 1934), 90.

States not only secured electricity's property form and regulated it as industry but also directly promoted electrification. Interconnected grids were overseen by national or regional dispatch stations, applying the technical and economic principles of load distribution in the name of national capital – which was at the time mostly in private hands.[86] State-led electrification went beyond the limits of short-term profitability and promoted power as a cheap article of mass production that should be available to everyone. This pushed the use of working nature into areas that had been formerly neglected, as in rural areas, agriculture, and poorer households. State-directed expansion was based on the understanding that electricity had become more than one form of power among others but was rather the very medium of economic development.[87]

Electrification as Socialization of Nature's Work

By the interwar years, the electric power industry had become highly concentrated and interconnected. Companies had merged, pooled their patents, coordinated prices, formed trusts, and cooperated in a wide range

86 Lagendijk, *Electrifying Europe*, 114.

87 Out of fear of lagging behind the US and Germany in electrotechnical development, the British government set up a Central Electricity Board in 1926 to press ahead with grid interconnection. The board's policy was to pressure supply companies explicitly in order 'to win incremental loads at cost'; since interconnection would eventually lower the costs of production, prices 'thus reflected their view of the long-run costs, not the historic accounting costs of public supply'; Hannah, *Electricity Before Nationalisation*, 179. Another example of this dynamic was the Tennessee Valley Authority (TVA), one of the most prominent cases of national electrification: It was a US development programme dating to 1933 that sought to use the power of the Tennessee River to electrify and irrigate nearby regions and sold electricity at production cost – its progenitors were proud of this 'leverage' it held over its private competitors. See, for the history of the TVA, Philip Selznick, *TVA and the Grass Roots: A Study of Politics and Organization* (Quid Pro, LLC, 2011); David Ekbladh, '"Mr. TVA": Grass-Roots Development, David Lilienthal, and the Rise and Fall of the Tennessee Valley Authority as a Symbol for U.S. Overseas Development, 1933–1973', *Diplomatic History* 26, no. 3 (1 July 2002): 335–74.

of ways. A few leading corporations were invested in the production of the entire gamut of machines and devices required for industry, utilities, and households, as well as in a range of other energy services.[88] Electricity not only brought light but powered motors, streetcars, and all sorts of tools used across society. The supply of electricity no longer was a manufacturing branch but had grown into something quite distinct. 'Electricity is not an industry,' noted Walther Rathenau, heir of the Allgemeine Elektrizitätsgesellschaft (AEG) and head of Elektrobank, in 1909,

> but a complex of industry, a sphere of the economy. It includes lighting and power, transport undertakings, the machine tool industry, the steam engine industry, centralisation and decentralisation of the production and distribution of power, and it links up with nearly all the other well-known fields of industry today.[89]

The expanding and all-encompassing nature of their business predisposed electrical engineers and managers to seize the influence that came with their position, to engage in broader social and political matters, and to offer 'theories' of electricity-based economic development. Through public office, talks, and books, some of them gained public prominence and sought to explain and promote electricity as a medium of development and socialization. This electrotechnical worldview was shared across national and political boundaries and not limited to engineers – social reformers of all stripes reflected on the social transformations triggered by electrification.

Electrification was understood as a process of socialization of nature and society. It meant to transform the unruly forces of nature into a controllable 'causality' (to recall Bogdanov's and Bachelard's phrasing), a force that could be used to transform the environment. At the same time,

88 Ronald W. Schatz, *The Electrical Workers: A History of Labor at General Electric and Westinghouse, 1923–1960* (University of Illinois Press, 1983), 5, 13; Stier, *Staat und Strom*, 44; Hausman, Hertner, and Wilkins, *Global Electrification*; Neufeld, *Selling Power*, 97.

89 Cited in James Joll, *Intellectuals in Politics: Three Biographical Essays* (Weidenfeld and Nicolson, 1960), 73.

it established new social relationships, mediated by electricity. Through electrical systems, people's lives had begun to depend on each other in a new way.

One of the most striking accounts of what electricity meant for the relation between humans and nature came from the British socialist and member of the Fabian Society Fred Henderson. In *Economic Consequences of Power Production* (1931), he described the pre-electric age as one in which human beings had to trick nature into doing work for them.

> [Man] could only use these other forces when he found them manifesting themselves in Nature on their own account. He could not command their presence at his will when and where he wanted them. When the wind died down or the stream ran low, his sails hung motionless, his ships lay becalmed and helpless, his wheels stopped turning. He could set his little trap across the track of this or that natural energy, and by waiting until it moved of its own accord along its own paths, he could borrow from it the movement which he wanted.[90]

This echoed a long-standing tradition to understand human labour as outwitting nature. It reads like a variation of Hegel's understanding of the working subject as 'lurking' for a self-moving nature to coincide with the subject's goals, where 'nothing happens to nature itself; the particular purposes of natural being become a universal purpose. The bird flies thither.' Yet Henderson suggested that this concept of labour was no longer adequate in the electric age, as 'man' could now make nature into pursuing its human purposes.

> Such casual and dependent use of a power which happens to be disporting itself regardless of human intentions is not the mastery of power on which modern industry has been created. To generate, to make energy appear and work at his bidding at his own chosen time and place, only when science taught Man that secret was power

90 Fred Henderson, *The Economic Consequences of Power Production* (G. Allen & Unwin, 1931), 17–18.

> dependably enslaved to human uses with the real mastery on which a system of production could be founded for the assured and regular service of the daily needs of human life.[91]

Nature's enslavement was a particularly strong metaphor in the Anglo-Saxon and French discourses.[92] While wind and water power could be used only for as long as nature would allow it, steam and even more so electricity were social creations. Henderson expressed this socialization of nature in a metaphor of ownership: Rather than *borrowing* a movement from nature, nature's power had now become man's property in the full legal sense of the word – it was as disposable and available as slave labour. Electricity, in this understanding, was the perfection of a human control over nature's work that had begun with steam.

Yet the mastery of nature through electricity could also take on more harmonious tones. Some highlighted that electrification was no perfection of steam but an entirely different natural force. When levelled back at nature, it could coordinate processes, mediate the conversion of forms of energies into each other, and change matter at a molecular level. Due to these properties, electrification could even heal the wounds of the first Industrial Revolution. This was expressed in the distinction, first formulated by Patrick Geddes, between a paleotechnic order based on steam and coal and the neotechnic order based on electricity, which would finally have 'its skill directed by life towards life, and for life'.[93] In the view of Gleb Krzhizhanovskii, head of the Russian electrification commission, electricity was a technology that could overcome the antagonism between nature and society: 'Having risen in a long and harsh struggle with nature . . .' he wrote in the popular journal *Bolshevik* in 1933, 'man finally senses the ways in which the powerful creation of his hands can be included in nature merely as an element that ennobles

91 Ibid.

92 Bob Johnson, 'Energy Slaves: Carbon Technologies, Climate Change, and the Stratified History of the Fossil Economy', *American Quarterly* 68, no. 4 (2016): 955–79.

93 Patrick Geddes, 'The Twofold Aspect of the Industrial Age: Paleotechnic and Neotechnic', *Town Planning Review* 3, no. 3 (1912): 181.

it.'[94] Similarly, David Lilienthal, director of the New Deal–funded Tennessee Valley Authority (TVA), spoke of a 'seamless web' of working nature linking human beings to land, water, and each other. Resources were now to be developed 'in that unity with which nature herself regards her resources'.[95] While these positions have to be seen in the context of the promotion and justification of the electrification projects they served, they were not entirely ideological: Electrification of production could conserve resources; electrochemistry could now isolate the elements, which capitalist agriculture had taken from the soil; and, with hydroelectricity, industrial production could again be distributed more equally over space.[96]

For electrotechnical thinkers, this new relationship to nature also implied changes in social relations that were already visible but not fully realized. For many contemporaries, the changes they witnessed in production processes and the economy as a whole seemed to gravitate towards some kind of socialization: There was a tendency in electricity that appeared to be overcoming the provision of power, and indeed all energy products, by individual units. Due to its business of building networks in cities, the industry had from the beginning traversed the boundaries of properties and involved some kind of public oversight. With hydroelectricity tapping into a public resource – water – and electric grids spanning entire regions, electric development by private companies now necessarily involved the state. The large interconnected systems could now cover the territory of entire nation-states (or more) and universalize the provision of power over time and space. Economic policies (a decision to invest in the steel industry or build a new electrochemical plant) as well as the ups and downs of the world market

94 Gleb M. Krzhizhanovskii, 'Marks o revolucionnom progresse tekhniki pri socializme', in *Sochineniia*, vol. 3, *Socialistichekogo stroitelstvo* (Akademia Nauk, 1936), 347.

95 David E. Lilienthal, *TVA: Democracy on the March* (Harper and Bros., 1944), 48.

96 This refers to an older process of nitrogen fixation from the air. The main process to produce fertilizers today is 'steam-reforming', which uses natural gas as a feedstock; Vaclav Smil, *Enriching the Earth: Fritz Haber, Carl Bosch, and the Transformation of World Food Production* (MIT Press, 2004), 39, 53–5.

(a surge in aluminium prices) affected their industry directly or indirectly. In turn, large electrification projects could not be realized without political involvement and a forecast of the future of general industrial production.[97]

Electricity also changed social relations in the factory. While steam power could reach the most distant corners of the world, its real spatial limit was the small scale. In steam-powered factories, the massive steam engine in the middle determined how materials could be routed and how machineries could be arranged in space. All tools driven by steam had to be connected through belts and pulleys, causing a lot of friction. Before production could begin, boilers had to be preheated and steam prepared. Whether all tools worked or not, the steam engine could not be easily regulated but had to be kept running at a certain level. Electricity allowed for a better utilization of capital, but not because it made power necessarily cheaper.[98] Its real value was that it could 'form' production differently.[99] According to the Soviet energy economist Veniamin Veic, through electricity, power could now be 'precisely applied, added, and split'.[100] The introduction of the electrical unit drive – each tool equipped with its own motor – could reveal the energy relations within a factory more accurately as each task was now energetically measured and stood, as electrical resistance, in a theoretically known quantitative relationship to all other units in the system.[101] In fact, only the electric motor can

97 Schatz, *The Electrical Workers*, 12.

98 Hannah, *Electricity Before Nationalisation*, 18. Electricity involved an additional conversion with the associated losses and the system was first modelled on the same system – the group drive – causing the same losses through friction; Warren D. Devine, 'From Shafts to Wires: Historical Perspective on Electrification', *Journal of Economic History* 43, no. 2 (June 1983): 357–61.

99 Devine, 'From Shafts to Wires', 371.

100 Veniamin I. Veic, 'Das Energieproblem in der gegenwärtigen Weltwirtschaft', in *Proceedings of the Second World Power Conference in Berlin, 1930*, vol. 16, *Allgemeine Probleme der Energiewirtschaft und gesetzlicher Fragen* (VDI Verlag, 1930), 33.

101 As a Siemens engineer reports: 'The electric motor can be used extremely well to determine the individual quantities of work of a transmission system, as it is only necessary to measure the amount of electrical energy supplied and to know the electrical resistances of the electric motor'; Ingr. Richter, 'Über elektrische Einzelantriebe', *Elektrotechnische Zeitschrift* 14, no. 10 (10 March 1893): 141.

measure the work performed by a machine precisely.[102] Electricity increased the flexibility of factory layout and enabled the controlled production lines and flow methods cherished in scientific management.[103]

As a coordinating force, electricity can automize power production. While Marx emphasized that the steam engine drew its own motive power, it did not haul its own coal or fire up its own boiler. With electricity, power ran by the turn of a switch, making some of the labour power around its production, transmission, and supply obsolete. Henderson described the working of such an automated power plant: From the delivery of coal to the switch board, the generation of 10,000 kilowatt-hours required the work of only half a dozen people. These workers 'touch the appropriate switch or button or lever at the appropriate moment'. They release or shut off power 'with an expenditure of [their] own physical energy . . . that would not lift a pound of coal three feet'.[104] Only

> two men . . . are required in attendance on this mammoth £50,000 power-machine. Getting the machine up to its full speed is a matter of about ten minutes at starting; opening the steam valve, starting the oil pump, and regulating the controls by adjustments to the gradual rise in speed until the machine is running all out. None of these operations involve more than small movements of the valve handles and the various levers of control. The human service required is solely concerned with the regulated admission and direction of power to the machine; and the acts of admission and direction are, generally speaking, the slightest touches on controls making no appreciable call for physical effort.[105]

Henderson's power plant was not typical of his time, but it was not science fiction either; it was entirely possible based on contemporary

102 Hermann Passavant, 'Mitteilungen aus dem Betriebe der Berliner Elektricitätswerke', *Elektrotechnische Zeitschrift* 16 (19 April 1894): 231.

103 Hannah, *Electricity Before Nationalisation*, 171.

104 Henderson, *The Economic Consequences of Power Production*, 24.

105 Ibid., 28.

technology. In this new automated world of production, the worker's physical energy was needed only in case the system broke down: Only 'at these outer fringes of work' did some of the old human labour survive.[106] Automatic operation could substantially reduce labour costs and meant that the pace of work was no longer regulated by the steam workers at the centre but was more firmly under control of the factory management.[107]

Once industry was integrated into a regional or national grid, the energetic forces of production were no longer owned by the industries that employed them. Electrification universalized energy access in time and space only through concentration of ownership. From many smaller manufacturers in the 1870s, the industry had now become dominated by a handful of large companies with hundreds of offsprings and patent relations between each other. German and American companies had almost entirely divided up the globe into spheres of electrical interests. This concentration, which we encountered as an explicit strategy in the previous part, was also noticed by contemporary observers. Meer Nochimson argued in *Die elektrotechnische Umwälzung* (The electrical revolution, 1910) – a book Lenin read in Berne – that this concentration manifests itself in terms of size and number of companies, as well as control over the energetic forces of production. Two years later, Kurt Heinig, the German Social Democrats' financial expert, gave a deeper account of the path of the German *Elektrotrust* in *Die neue Zeit*, detailing the many financial and legal relationships that integrated the German electro industry into only two powers, the Allgemeine Elektrizitätsgesellschaft (AEG) and the Siemens-Schuckert corporation.[108] In *Imperialism: The Highest Stage of Capitalism* (1916), Lenin calculated that

> less than one-hundredth [1 per cent] of the total number of enterprises utilise more than three-fourths [75 per cent] of the total amount of

106 Ibid., 29.

107 Hannah, *Electricity Before Nationalisation*, 171; Neufeld, *Selling Power*, 80–5; Hugh Quigley, *Electrical Power and National Progress* (G. Allen & Unwin, 1925), 21.

108 Kurt Heinig, 'Der Weg des Elektrotrusts', *Die neue Zeit* 2, no. 39 (1912): 474–85.

> steam and electric power! Two million nine hundred and seventy thousand small enterprises (employing up to five workers), constituting 91 per cent of the total, utilise only 7 per cent of the total amount of steam and electric power![109]

The electrotechnical industry seemed to best express the general capitalist tendency towards monopoly and concentration.

Some Soviet engineers took this to mean that capitalism had reached an energetic impasse – electricity would turn out to be capitalism's final achievement. Within capitalist relations of production, electricity's potential could not be fully realized: Either small utilities resisted concentration, squandering the productive power of electricity in their petty-minded competition, or all utilities ended up in the hands of a few corporations, repeating an endless cycle of overproduction and crisis. Within a socialist country, electricity could serve the socialization of production, integrating the peasant's electro-plough, the small motor, and the huge factory into the same grid and coordinating their work just like in a single country-wide factory.[110] Echoing Bogdanov, Krzhizhanovskii thought this apparatus would form a completely different subject of labour. A socialist energetics would topple the 'autocracy of engineers' and the 'aristocracy of machinery' which had hitherto dominated workers. Instead, in the power economy, living labour would stand 'above a machine not in the form of an individual creator of this machine – an engineer – but in the form of a conscious human collective armed with the creative thought of centuries'.[111]

This vision of electricity as a great social unifier was not limited to Soviet propaganda of the electrification plan. It was also not entirely

---

109 Vladimir Ilich Lenin, *Imperialism: The Highest Stage of Capitalism* (Penguin, 2010), 13–14.

110 Iosif Ivanov, 'Materialnyi bazis kommunisticheskogo obshchestva', *Vestnik socialisticheskoi akademii* 4 (1923): 175; Gleb M. Krzhizhanovskii, 'Elektrifikaciia kak faktor obobshchestvleniia proizvodstva', in *Sochineniia*, vol. 1, *Elektroenergetika*, ed. Akademia Nauk SSSR (Gosudarstvennoe Energeticheskoe Izdatel'stvo, 1933), 80–4.

111 Gleb M. Krzhizhanovskii, 'Energetika i socialisticheskaia rekonstrukciia', *Planovoe khoziaistvo* 1 (1929): 11.

unfounded, as interconnection and diversification rested on the idea of loads enabling each other in electrical systems: The cheap supply of a small, irregular load was possible as long as there were other, more steady load patterns. The worker's household could be served not despite but because of the capitalist's industrial load. This complementary nature of electrical systems strengthened the idea that a more harmonious and united society could be realized on this basis and that everyone's electric needs could be folded into a given power capacity over time.

The leading metaphor was now no longer the machine but the productive system or organism: Nature and society, workers, households, and industry, were all elements of a growing, developing organism. Influenced by Edward Bellamy's *Looking Backward*, Walther Rathenau imagined 'a community of production, in which all members interlock organically, united to the right and left, above and below, into a living unity endowed with uniform perception, judgment, force, and will; not a confederation, but an organism.'[112] While the concrete proposals differed, the large industrial and, above all, electrotechnical corporations served as a model to reorganize economic life. According to Rathenau, their methods of economic efficiency should be transferred to other sectors and the economy as a whole.[113] This was at least partly a utopia of depoliticization in a time of fierce political struggles. It was typical for the electrotechnical intellectuals to imagine themselves in a position separate from politics and industry, with their only obligation the full development of productive capacity. The metaphor of an efficiently regulated social organism left no room for politics or competition – its parts would neither compete nor quarrel but collaborate for the benefit of the organism as a whole.[114]

The electrotechnical thinkers were committed to an idea of emancipation through the rationalization and socialization of production. For some, the purpose was to reduce waste, luxury, and idleness

112 Walther Rathenau, *Die neue Wirtschaft* (S. Fischer Verlag, 1918), 235.

113 Adolf Günther, 'Walther Rathenau und die gemeinwirtschaftlichen Theorien der Gegenwart', *Weltwirtschaftliches Archiv* 15 (1919/20): 44.

114 John M. Jordan, '"Society Improved the Way You Can Improve a Dynamo": Charles P. Steinmetz and the Politics of Efficiency', *Technology and Culture* 30, no. 1 (1989): 57–82.

– emancipation was the insight into the relations of social reproduction. In Rathenau's 'Common Economy' (*Gemeinwirtschaft*), associated professions, similar to guilds, would oversee and coordinate rational production.[115] Similarly, Owen D. Young, then chairman of the board of General Electric (into which Edison's company had merged), envisioned a worker-controlled corporation and a classless society in which everyone would feel personally responsible for full production:

> I hope the day may come when these great business organizations will truly belong to the men who are giving their lives and their efforts to them . . . Then they will use capital truly as a tool and they will all be interested in working it to the highest economic advantage. Then an idle machine will mean to every man in the plant who sees it an unproductive charge against himself. Then every piece of material not in motion will mean to the man who sees it an unproductive charge against himself . . . Then we shall dispose, once and for all, of the charge that in industry organizations are autocratic and not democratic . . . Then, in a word . . . we shall have no hired men.[116]

This not only made matters of social reproduction into everyone's responsibility but also limited everyone's purpose to reproduction. As in an organism, the form and function of each part were to reflect its relation to the whole, as every part of society would understand itself as essential to the growth of the entire system. Human society meant, above all, participation in social and material reproduction. Several socialist thinkers, however, argued that the purpose of rationalized production could only be emancipation from work.[117] Both groups,

115 Hans Dieter Hellige, 'Walther Rathenaus Pionierrolle in den Diskursen über das Nachhaltigkeitsproblem der Moderne', in Sven Brömsel, Patrick Küppers, and Clemens Reichhold, eds, *Walther Rathenau im Netzwerk der Moderne* (De Gruyter, 2014), 136–85; Günther, 'Walther Rathenau und die gemeinwirtschaftlichen Theorien der Gegenwart', 52–3.

116 Owen D. Young, 'Dedication Address', *Harvard Business Review* 5, no. 4 (1927): 392.

117 Notably Charles Steinmetz, Edward Bellamy, and Carl Ballod; see N. W. Balabkins, 'Der Zukunftsstaat: Carl Ballod's Vision of a Leisure-Oriented Socialism',

however, joined the emancipation of the working class to the exploitation of working nature.[118]

Electricity assumed a new meaning in the imaginary of socialized production. It was no longer the expression of a 'force' realizing itself in animate and inanimate objects, as imagined in natural philosophy. Nor was it a potential stored by nature to fuel humanity's machines, as in the steam economy. Rather, those parts of the body used as metaphors for electrical power expressed the new structuring function of working nature: 'Power constitutes the life-blood of industry in the modern state', wrote Hugh Quigley in *Electrical Power and National Progress* in 1925, 'it makes labour capable of performing its functions in production; it lays the foundation for all developments, financial and technical, and renders it possible for man to derive from nature the utmost wealth nature is capable of yielding.'[119] Krzhizhanovskii spoke of power as the 'backbone' of the society the Bolsheviks were creating.[120] Without blood, the body quickly deteriorates; without a backbone, it loses all shape. In a distinct change of the social and the natural, the socially crafted currents now sustained the organism of the social: Humanity governed its own life conditions. In its most technical form, working nature had become an organic part of society.

The electric current is a social creation, but it is not created under freely chosen conditions. To make electrons drift in a controlled way requires a conductor of a particular material (and a non-conductor of other materials), a circuit of certain dimensions and specifications (voltage, current, and resistance must relate in a certain way), and a voltage source such as a battery or generator (which, in turn, requires an external source of power). In some languages, the electrotechnical vocabulary that

---

History of Political Economy 10, no. 2 (1 June 1978), 213–32; Carl Ballod, *Der Zukunftsstaat: Produktion und Konsum im Sozialstaat* (Dietz, 1919); Jordan, '"Society Improved the Way You Can Improve a Dynamo"'.

118 Walter Benjamin, *Illuminations: Essays and Reflections* (Schocken, 1997), 259.

119 Quigley, *Electrical Power and National Progress*, 16–17.

120 Gosplan, 'Plenum Gosplana', *Planovoe khoziaistvo* 6–7 (1923): 46.

captures these conditions still reflects the field's dialogue with natural philosophy. The German words for 'voltage' (*Spannung*, or tension) and 'resistance' (*Widerstand*) were important concepts in Friedrich Schelling's attempt to understand how opposing forces organized matter. Electric phenomenon, a unity created by polarization, manifested a reconciliation of forces whose former identity expresses itself in a tension that is directed at each other. Polarity thus reminisces a former unity and signals identity, not non-identity. 'Where a natural force meets resistance, it forms its own peculiar sphere, the product of its intensity and the resistance it finds.'[121] Resistance, in this understanding, is productive and brings about a new configuration.

In the management of the current, electricity's resistance to its creation and exploitation can be said to have been productive in this way. While the peculiar materiality of the electric current limited the early lighting systems, it soon served to guide the management of the current as an energy commodity. The electricity business constitutes a further moment in the subsumption of the prime mover, as it enabled the production of a certain capacity to perform work over time – and, ultimately, a further centralization of the means of energy production. To be profitably managed, those systems had to be shielded from competition. They needed access to a region's entire diversity of social life, which – with the right tariff incentives – could be transformed into a well-distributed electricity demand. While such systems still employed prime movers in the form of turbines and oscillating masses, subsumption now superseded this narrow meaning of mechanical work and went beyond the control of heat and motion: Electricity could be employed to create matter, transmit sound, and store information.

Like steam, on which it still rested, this power economy generalized the use of 'working nature' and engendered its own spatial and temporal order. In space, production became even less dependent on concrete marks of the landscape, as electricity could be generated from a wider range of sources. It revived the use of water and made, through

121 Friedrich Wilhelm Schelling, *Von der Weltseele* (Friedrich Werther, 1806), 26.

interconnection and a higher level of consumption, even more secluded streams profitable. It also drew new types of fuel, peat and lignite, into the firebox. Power transmission enabled a spatial distance between generation and consumption but also posed a new limit to the area over which electricity could be supplied economically. What is more, it made power scalable and therefore applicable at a higher spatial resolution. Gretchen Bakke argued that electricity was a 'being in time' not in space and that the electric grid was managed as a temporal rather than a spatial order.[122] As the electrical field materializes and changes at the speed of light, electric systems come with a punctuated temporality – electricity must be consumed or absorbed by a resistance within a millisecond of its generation. This constituted the material basis for the corporate strategies of folding ever more consumers and users into the system's capacity over time. For all the improvements to grid safety, techniques of self-regulation, and the means of storing surplus electricity in potential or chemical energy, this basic principle still applies, down to the present.

Within this most objectified form of energy, the problem of putting nature to work appears again on a different level. Electricity does not so much supersede as mediate concrete space and time: The natural moment comes to matter differently. Resistance still limits transmission, and climate, weather, seasons, holidays, and working days appear in the records of power stations as changes in the patterns of electricity generation, demand, and variations in capital utilization. With the integration of wind and solar power, systems will have to take into account weather patterns at a time when they might already be mediated through climate change.

Electrification, an unconscious socialization of ever more activities through electricity, is still ongoing. But the hopes, voiced in the interwar years, for a more consciously coordinated use of nature's work have amounted to little more than nationalization (in some countries) and public regulation of private monopolies (in others). Both forms reflected the conviction that electricity had become a public good and that its most efficient organization was one that restricted competition. By keeping electricity cheap in the postwar decades, these regulations also

122 Bakke, 'Electricity Is Not a Noun', 33.

quelled calls for deeper socialization. When electricity finally became politicized again in the 1970s, the focus was on decentralization and the right to self-produce electricity from other sources, mostly wind, which the big companies were slow to adopt.[123] Ironically, the integration of renewables can be seen as the first step in breaking up monopolies, and their rise coincides with the creation of electricity markets in the 1990s.[124] Since then, the dominant paradigm has been socialization through the market – a paradigm challenged rhetorically, but not practically, by calls for decentralized self-production. The idea that electricity, because of the socialized form in which it can only be realized, has a unique potential to mediate social relations has given way to a more liberal understanding of citizens being energetically independent from each other. However, as the task of making electricity permanent by basing it on renewables increasingly runs against the financial industry's expected returns on investment, the Soviet engineer's conviction that electricity was capitalism's last achievement may still turn out to be true.[125]

---

123 Richard F. Hirsh, *Power Loss: The Origins of Deregulation and Restructuring in the American Electric Utility System* (MIT Press, 1999).

124 Matt Huber, 'Unbundling the Grid: Renewables Capital and the Demise of Electricity as a "Public Utility" in the United States', *Development and Change*, 7 August 2025, 982–1006; Ronan Bolton, *Making Energy Markets: The Origins of Electricity Liberalisation in Europe* (Springer, 2021).

125 Brett Christophers, 'Fossilised Capital: Price and Profit in the Energy Transition', *New Political Economy* 27, no. 1 (2 January 2022): 146–59; Brett Christophers, 'Taking Renewables to Market: Prospects for the After-Subsidy Energy Transition: The 2021 Antipode RGS-IBG Lecture', *Antipode* 54, no. 5 (19 May 2022): 1519–44; Brett Christophers, *The Price Is Wrong: Why Capitalism Won't Save the Planet* (Verso, 2024).

5

Governing Energy

Fuel joules should not be added to electrical joules (except in the very special situation when the interest is only centered on the aspect of heating the Earth as a whole).

– World Energy Conference and International Union of Producers and Distributors of Electrical Energy, *Substitutions Between Forms of Energy and How to Deal with Them Statistically: A Guide*

Today's climate models not only contain staggeringly complex equations of the earth's carbon cycle but also draw on more prosaic numbers, such as energy statistics. The total primary energy supply (TPES), as well as the rate at which carbon dioxide is produced in fossil fuel combustion, is crucial to estimate the trajectory of the planet's temperature and to distinguish natural from human-made climate change. Calculating mitigation pathways also relies on data on the 'energy-supply sector', as the Intergovernmental Panel on Climate Change (IPCC) calls the global energy economy. In fact, the earliest climatic models were simple energy balances, which treated the whole earth as a single system of incoming and outgoing energy; the human energy economy mediates what happens in between. Integrated assessment models (IAMs) are so called because they combine energy, environmental, and economic submodels. Even though energy data was at the core of these models from the beginning, their historical epistemology has received little attention. It constituted a relatively unspectacular and uncontentious input into an otherwise highly politicized model.[1]

This is not because making global energy data was self-evident. Quite the opposite, the attempt to make statistics of coal, oil, gas, water, and electricity comparable brought to light their deeper incomparability – as the quote above attests to. Like all highly aggregated statistical data, energy data requires well-equipped institutions and an entire statistical apparatus. To make numbers appear simple, to make them comparable, reliable, and understandable, takes a lot of work. Energy statistics are no exception. Their immediacy and simplicity of meaning are highly mediated. The ease with which one can today compare the relative energy supply of resources that are materially very diverse, undergo very different technical processes, and are used for various purposes at different places around the world is the result of decades of statistical and standardizing work. As a basic form of market information, such numbers are involved in all kinds of economic decisions – about whether to expand an energy business, how much to produce, or whether it is worth investing in a certain energy sector of a certain country. But they are also the material from which the arguments in today's debates on resource justice, energy transitions, and climate change are forged.

This chapter traces the history of the method that made working nature economically comparable – energy balances – from early attempts in the 1930s to the first global energy balances in the 1980s. The UN's *International Recommendations for Energy Statistics* (2018) defines energy balances as an 'accounting framework for compilation of data on all energy products entering, exiting, and used within the national

1 Paul N. Edwards, *A Vast Machine: Computer Models, Climate Data, and the Politics of Global Warming, Infrastructures* (MIT Press, 2010); Matthias Heymann and Amy Dahan Dalmedico, 'Epistemology and Politics in Earth System Modeling: Historical Perspectives', *Journal of Advances in Modeling Earth Systems* 11, no. 5 (2019): 1139–52.

territory of a given country during a reference period'.[2] The basic form of an energy balance is a table, where the columns represent various forms of energy, and the rows show their way from production to consumption. In other words, balances are a way of combining production and consumption statistics of different resources to show the flow of 'energy' and its conversions through an economic space. Within this framework, all energy products are related to each other: As different as raw oil, solar power, and heating gas might be in their material appearance and utilization, the model brings them into a precise and fixed quantitative relation – they become equivalent in their ability to perform work.

In most OECD countries, which are the focus of this chapter, this statistical work began during economic reconstruction and the politics of economic growth after World War II. Governments and companies sought to rationalize and justify their investment decisions when the coal, oil, and electricity industries became increasingly interrelated. The balance has long been an instrument to govern resource flows vital to state power, such as grain and money. Energy balances enter corporative and national statistics as a means to deal with the increasing competition between energy commodities, when it became technically and economically possible to produce forms of energy from a range of sources – heat from coal and petroleum, electricity from water, fuels, and nuclear. Globally, this space for substitution develops very unevenly. However, by the 1950s and 1960s, comparisons between energy products became necessary for a wider range of actors. Managers, engineers, and politicians in the OECD countries and beyond agreed that energy resources must be treated collectively, that their interrelation should be formalized, and that energy balances are the method to do so. The statistical abstraction of energy follows from this imperative to compare the costs of units of energy and the profitability of one source compared to another.

Energy balances are an abstract model of the energy economy to render negotiable and governable the existing and future relations between resources that can perform work. Their success has been

2 United Nations Statistical Division, *International Recommendations for Energy Statistics (IRES)* (United Nations, 2018), 17.

possible only because the many ways in which forms of energies are different, and incomparable in their economic function and social meaning, are disregarded. As a model, they can be used to 'represent' the existing energy economy in a way that highlights the energetic relations between forms of energy. The quote that started this chapter illustrates that this always implies decisions on what counts as an energetic process and how they should be valued: If the interest is on the 'useful work' resources can do, rather than on their ability to 'heat the Earth as a whole', fuel joules and electric joules are not identical. In formalizing equivalence, energy balances also go beyond existing relations: They open up the horizon of each unit of energy, anywhere in the world, being related to any other. By way of establishing *theoretical* relations between *all* forms of commodified energy, and institutionalizing them, balances enable the imagination of connections that do not yet exist or demands that could be satisfied.

Through the common problem of 'oil substitution' in the 1970s, the OECD's balancing framework became the basis for national energy balances around the world and for the compilation of a world energy balance. Following slightly different methodologies, global energy balances are today routinely compiled and published by the United Nations Statistical Department, British Petroleum, the International Energy Agency (IEA), and the US Energy Information Administration (EIA). Thus, when the IPCC was eventually founded in 1988, there was already an international infrastructure in place to gather standardized data on all forms of energy, individually and collectively. Today, the IPCC still relies on IEA data, which, in turn, relies on data from governments and industry.[3] As tables or flow diagrams, energy balances have become ubiquitous in the field of energy, climate, or environmental policy.

3 Groupe d'Experts Intergouvernemental sur l'Évolution du Climat, eds, *Climate Change 2014: Mitigation of Climate Change, Working Group III, Contribution to the Fifth Assessment Report of the Intergovernmental Panel on Climate Change* (Cambridge University Press, 2014), 1293–5.

A New Factor of Production?

Amid a deteriorating balance-of-payment crisis during the Great Depression, Hungarian chemical engineer Ernö Haidegger presented an energy balance of Central Europe at the World Power Conference in Berlin in 1930. He pointed to an interdependence of a new kind: Countries were connected through relations of energy, just like they were through relations of capital. This was especially true for the countries of Central Europe, most of which had emerged out of the fall of the Habsburg Empire and which Haidegger treated as an economically integrated whole. The war had capped trade relations, broken apart empires, and thereby created a range of small countries, political 'stumps' whose economic and energetic viability was unsure. Haidegger's balance could well be the first energy balance that went beyond a national territory and showed to what extent states were energetically dependent on each other. The balance, stitched together from import-export and production data, showed how some countries were energy self-sufficient, while others depended heavily on imports.[4]

Unlike some of the electric intellectuals we met in the previous chapter, Haidegger did not believe in a deeper energetic reality of the economy. Energy accounting would not make monetary accounting obsolete; it was only truly revealing when paired with a balance of payments for energy products. By doing so, one could see that, even though Central Europe appeared self-sufficient in terms of energy, meaning that the total energy consumed roughly equalled the energy produced, energy imports still affected the balance of payments negatively. The need for energy was a drain on Europe's capital. This fact was easily explained, however, 'if you consider that the continent's export is limited to coal while its import primarily consists of oil', a market dominated by American companies. Thus, even in this field, 'the European continent owes a couple of Million Marks annually to the mighty America'.[5] Haidegger's paper already brought

4 E. Haidegger, 'Die Energiebilanz und ihre Gestaltung in den Staaten Mitteleuropas', in World Power Conference, ed., *The Transactions of the Second World Power Conference*, vol. 4 (VDI Verlag, 1930), 27.

5 Ibid., 28.

up the questions of energy and capital that are at the heart of energy balancing.

Following his talk, Haidegger called for the World Power Conference (WPC) to include energy balances in the organization's statistical programme. The WPC had been founded in 1924 by Daniel Dunlop, a British electrotechnical engineer with anthroposophical leanings. In the postwar spirit of technical internationalism, it brought together the engineers, scientists, and economists concerned with fuel and power that industrialization had bred. Within the tense political atmosphere, the conference was to act as 'nothing more than a clearing house for the interchange of information on all matters relating to the development of power resources'.[6] The financial metaphor is telling: Just like a clearing house in banking facilitated exchange between buyer and seller by overseeing and standardizing the payment process, the WPC was to cater to producers and consumers of power and aid the circulation of resources and technologies by collecting and sharing information. After the first conference, the WPC began to publish data on power resources, production, and consumption regularly.[7]

Collection and aggregation of national resource statistics had become geopolitical arm wrestling at the time.[8] The World Coal Congress had

6 Electrical World (1925) cited in Rebecca Wright, Hiroki Shin, and Frank Trentmann, *From World Power Conference to World Energy Council: 90 Years of Energy Cooperation, 1923–2013* (World Energy Council, 2013), 16.

7 Johan Schot and Vincent Lagendijk, 'Technocratic Internationalism in the Interwar Years: Building Europe on Motorways and Electricity Networks', *Journal of Modern European History* 6, no. 2 (2008): 196–217; Daniela Russ, 'Speaking for the "World Power Economy": Electricity, Energo-Materialist Economics, and the World Energy Council (1924–78)', *Journal of Global History* 15, no. 2 (2020): 311–29.

8 Andrea Westermann, 'Geology and World Politics: Mineral Resource Appraisals as Tools of Geopolitical Calculation, 1919–1939', *Historical Social Research* 40, no. 2 (2015): 151–73; Andrea Westermann, 'Inventuren der Erde: Vorratsschätzungen für mineralische Rohstoffe und die Etablierung der Ressourcenökonomie', *Berichte zur Wissenschaftsgeschichte* 37, no. 1 (March 2014): 20–40; Daniela Russ and Thomas Turnbull, 'Competing Powers: Engineers, Energetic Productivism, and the End of Empire', in Daniela Russ and James Stafford, eds, *Competition in World Politics* (transcript, 2021), 183–210.

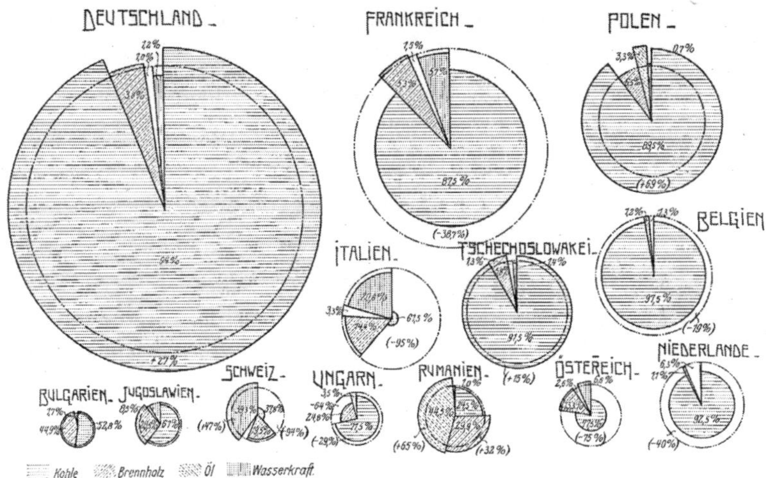

Figure 5.1. The structure of Central European energy economies, published in Ernö Haidegger, 'Die Energiebilanz und ihre Gestaltung in den Staaten Mitteleuropas', in World Power Conference, ed., *The Transactions of the Second World Power Conference*, vol. 4 (VDI Verlag, 1930), 27. Overflowing circles indicate a positive balance, empty circles a negative balance.

published its *The Coal Resources of the World* in 1913 and the League of Nations compiled world resource statistics in the 1920s as well. Oil, coal, and water power statistics were often added together by making use of the theoretical equivalence stated by the law of energy conservation. The WPC's *Power Resources of the World (Potential and Developed)* of 1929 did just that: It ends with a table on world production, a recalculation of single resources on a common basis, 'so that it is possible to understand very clearly the part played by each source of energy, the percentage of world power production represented by water-power, for example, as compared with coal'.[9] What this 'part' meant in practice, however, was not at all clear. It was easy to express water power and coal resources in British thermal units by recalculating them in units of work, making use of the equivalence stated by physics, but

9 Hugh Quigley and D. N. Dunlop, *Power Resources of the World (Potential and Developed)* (World Power Conference, 1929), vii.

it was impossible *to make them do the same work.* Not all power resources could replace each other in industry: Electricity could not power ships. The conversion was more easily performed in theory than in practice. The concept of energy in physics spanned a universal horizon of equivalence that went beyond what was technically possible: Everything could be described in terms of energy, but not everything could be converted into each other, let alone converted in a way that work was performed.[10]

A balance went beyond the aggregation of resource statistics and sought to express real economic relations. This made sense for commodities that circulated widely through the economic space, affected all economic activities, and were critical to state power. This was typically the case for grain or capital.[11] Haidegger argued that energy had just become such a commodity; in fact, it had become a factor of production like labour, land, and capital. Replacing land by nature, Haidegger pointed out that 'nature actually contributes to production in two forms, and those two factors are: material and power [*Kraft*], or, rather, energy'.[12] The latter, he added, had

10 'Power' was a subtle resource that resisted the way in which other riches were measured. It appeared in confusingly diverse forms. An early way to communicate this hidden identity to a wider public was the use of 'coal' itself as a metaphor for power. The term 'white coal' for hydroelectricity perhaps illustrates best how industrial interests and imperial competition jointly drove these energetic comparisons: Popularized by the French industrialist Aristide Bergès in the 1870s and '80s, the term invoked the metaphorical space of coal – the fuel of the most powerful empire, Great Britain – to emphasize that similar potentials lay hidden under the French territory: The glaciers of the Alps could be mined just like British or German coal seams. ; Tobias Silseth, 'Energy and Ideas of Progress in France and Britain, 1865–1924' (PhD thesis, University of Cambridge, 2024), 174–84. The metaphor circulated widely and helped to 'read' the national space as an energetic landscape before it was measured as such. Taking it to its somewhat silly extreme, forty years later the geophysicist Boris Veinberg published books on coals of different colours – including on labour as 'red coal' – and a summary entitled 'Coal – Red, Black, Yellow, Blue'; Boris Petrovich Veinberg, *Ugol' chernyi, krasnyi, zheltyi, belyi* (Nauchnoe Knigoizdatelstvo, 1925).

11 Marina Fischer-Kowalski, 'Society's Metabolism: The Intellectual History of Materials Flow Analysis, Part I, 1860–1970', *Journal of Industrial Ecology* 2, no. 1 (1 January 1998): 61–78; Helmut Haberl, 'The Energetic Metabolism of Societies Part I: Accounting Concepts', *Journal of Industrial Ecology* 5, no. 1 (1 January 2001): 11–33.

12 Haidegger, 'Die Energiebilanz und ihre Gestaltung', 3.

gained a particular significance and constituted 'the most important and irreplaceable factor for modern economic production.'[13] Haidegger was not alone in thinking that power had by then become a general condition of production. Work along these lines was done at the Brookings Institution in the US, where Frederick Tryon had turned to the question of energy and economic output; at the Socialist Academy of Science in Moscow, where Maria Falkner-Smit studied the relation between productivity and power; and probably in other places around the world.[14] Not only did this imply a degree of substitutability between the factors of production – labour, capital, and energy – it also meant to treat various resources which underwent different technical processes *as yielding the same productive effect* regardless of time and space.

The balance method has a long history in the form of double-entry bookkeeping among early capitalist merchants and industrialists. François Quesnay was the first to represent the flow of payments across different sectors of the economy in tabular form. Underlying this method was the idea that there is an economic cycle of reproduction that could be represented by measuring what each sector received from and returned to each other. Beyond a mere aggregate of national income that preceded the gross domestic product, a balance could show the creation and metamorphosis of social capital. Over the nineteenth century, this idea was picked up by Léon Walras and Karl Marx, among others, and the latter gave his own version of it that allowed him to distinguish between 'simple' and 'expanded reproduction', between an economy that reproduced itself in perpetual identity and a growing economy that transformed itself.[15]

13 Ibid.

14 Maria Falkner-Smit, 'The Motor Power Outfit of Labour and Its Economic Efficiency', in *The Transactions of the Second World Power Conference*, vol. 16, *World Problems of Power Economics* (VDI Verlag, 1930), 60–71; Antoine Missemer and Franck Nadaud, 'Energy as a Factor of Production: Historical Roots in the American Institutionalist Context', *Energy Economics* 86 (February 2020): 104706; F. G. Tryon, 'An Index of Consumption of Fuels and Water Power', *Journal of the American Statistical Association* 22, no. 159 (September 1927): 271–82.

15 Karl Marx, *Capital. A Critique of Political Economy*, vol. 2 (Penguin, 1993), 435–6.

The method returned when states tried to manage economic life more closely amid the economic turbulences of the interwar years. Wassily Leontief, who would later receive the Noble Prize for his work on 'input-output analysis', was influenced by the attempt of the Soviet planning bureaucracy to set up a national economic balance in the mid-1920s. The Soviet balance, calculated from very poor data by the Central Statistical Office in 1926, took explicit inspiration from Quesnay and Marx's table of economic reproduction. It included a simple energy balance, which was deemed vital to plan expanded economic reproduction.[16] In the 1930s, inspired by Wagemann's *Konjunkturlehre*, German statisticians began to use their statistical data on industrial production in similar ways, making it possible 'to trace the economic interrelationships between the sectors of industry and the relations between industry and the rest of the economy'.[17]

Balances required a new type of statistical input: data on what and how much was produced from which input materials. Industrial censuses had been pioneered in the UK and the US in the early twentieth century, but only the more detailed surveys could serve as raw material of a balance.[18] The British census of production asked firms to 'declare the value of their output and the total value of all materials' and labour they bought, yielding a picture of the net value of production at each point.[19] The US survey was more comprehensive and required companies to itemize the raw material they used and give their output in terms of both physical quantities and value. While such production surveys were retrospective, they could in principle be used for the forward planning of production – just as socialist

16 Wassily Leontief, 'Die Bilanz der russischen Volkswirtschaft: Eine methodologische Untersuchung', *Weltwirtschaftliches Archiv: Zeitschrift für allgemeine und spezielle Weltwirtschaftslehre* 22, no. 2 (1925): 338–44; P. I. Popov, 'Introduction to the Balance of the National Economy', in Nicolas Spulber, ed., *Foundations of Soviet Strategy for Economic Growth* (Indiana University Press, 1964), 5–19.

17 Adam Tooze, *Statistics and the German State, 1900–1945: The Making of Modern Economic Knowledge* (Cambridge University Press, 2001), 200.

18 Frederick G. Bohme, 'U.S. Economic Censuses, 1810 to the Present', in 'Symposium on the Economic Censuses, United States Bureau of the Census', special issue, *Government Information Quarterly* 4, no. 3 (1 January 1987): 221–43.

19 Tooze, *Statistics and the German State*, 191.

countries later did with material balances.[20] Within these production statistics, however, the output of the coal, oil, and electric industries appeared as individual sectors with their own respective units – kilowatt-hours in electricity, British thermal units in heat processes, weight or calories for fuels, and so on – which could not be readily compared.

One site where *energetic* balancing had taken place before was in the thermodynamic optimization of engines and factories. Thermodynamics had developed around the study of steam engines. However, the insights in theoretical thermodynamics by scientists like Gustave Adolph Hirn, William Thomson, or Rudolf Clausius only slowly found their way back into engine building, facilitated by the new and more efficient steam engines that were required for ocean navigation and electric systems in the late nineteenth century.[21] Henry Riall Sankey, an Irish-born captain in the Royal Navy who was charged to find a cheap way to generate electricity, began working on the thermal efficiency of steam engines in the 1890s. Heading a commission on the matter, he presented a report on the standards of thermal efficiency to the British Institution of Civil Engineering in 1898. While Sankey is now best known for the flow diagram in which he expressed the movement of heat through the steam plant, the illustration was only a by-product of his work on conceptualizing an ideal steam engine – one generating the highest theoretically possible amount of work – against which existing steam engines could be measured.[22] Over the following

20 John Bennett, *The Economic Theory of Central Planning* (Blackwell Publishers, 1989); J. M. Montias, 'Planning with Material Balances in Soviet-Type Economies', *American Economic Review* 49, no. 5 (1 December 1959): 963–85; Tooze, *Statistics and the German State*, 191–207.

21 D. S. L. Cardwell, 'Steam Engine Theory in the 19th Century: From Duty to Thermal Efficiency; From Parkes to Sankey', *Transactions of the Newcomen Society* 65, no. 1 (1993): 121–2.

22 H. R. Sankey, 'The Thermal Efficiency of Steam Engines', *Minutes of the Proceedings of the Institution of Civil Engineers* 125 (1896): 182–212; H. R. Sankey, 'The Thermal Efficiency of Steam Engines: Report of the Committee Appointed to the Council upon the Subject of the Definition of a Standard or Standards of Thermal Efficiency for Steam Engines', *Minutes of the Proceedings of the Institution of Civil Engineers* 134 (1898): 278–312; Mario Schmidt, 'The Sankey Diagram in Energy and Material Flow Management', *Journal of Industrial Ecology* 12, no. 1 (2008): 82–94.

decades, energy and entropy balances were drawn for steam power plants and other industrial processes that involved heat as a means to identify losses and improve efficiency.[23]

National energy balances essentially treated the national economy as a single, differentiated energy-converting process. This breathtaking idea had shimmered through the metaphor of society as a 'productive organism', but it had never been operationalized. To do so, energy balances necessitate a recalculation of the individual input resources (coal, water power) and products (coke, electricity, etc.) into a common unit (Haidegger used the calorie). What sounds like a simple task is not easily done: Production statistics on fuels usually came in a unit of weight (tons of oil or coal). Because fuels performed very differently in combustion, they had to be weighted according to the amount of heat they could produce. As heating values, though widely discussed since the mid-nineteenth century, were at that time still not standardized across countries (see chapter 3), Haidegger simply used the conventional heating values of the German-speaking coal regions. But this was only to account for the differences in the energetic potential of the resources themselves, not for the differences in how this potential was realized in different technical processes. 'Strictly speaking,' Haidegger noted, 'the calorific valuation of fuels does not lead to an accurate result, as it does not account for the efficiency of the energy conversions taking place.'[24] A steam engine realized half as much of a coal's energetic potential as a combustion engine did of petroleum's. Though well known to engineers and discussed under the concept of 'useful work', this quantity did not exist as the result of any regular, statistical collection. Conversion factors beyond fuels were even more complicated. How could fuels and water power be treated as equals, given that they undergo such different

23 Marina Fischer-Kowalski and Walter Hüttler, 'Society's Metabolism: The Intellectual History of Materials Flow Analysis, Part II, 1970–1998', *Journal of Industrial Ecology* 2, no. 4 (1 October 1998): 107–36; Marco P. Vianna Franco, 'The Factual Nature of Resource Flow Accounting in the Calculation in Kind of the "Other Austrian Economics"', *Œconomia: History, Methodology, Philosophy* 10, no. 3 (1 September 2020): 453–72.

24 Haidegger, 'Die Energiebilanz und ihre Gestaltung', 20.

technical processes and become comparable only in the last step – as with electricity?[25]

Haidegger's proposal sparked a discussion on energy balancing among the WPC's national committees that ensued over the following years. While a few national committees claimed that balancing work was already underway in their statistical bureaus (among them the Soviet Union, Argentina, and South Africa), the majority saw themselves incapable of furnishing the necessary data. The main point, however, raised by many national committees, was the impossibility to decide theoretically on a global conversion factor between different forms of energy. But there were also more general objections. Oscar C. Miller of the US Federal Power Commission denounced this kind of accounting as 'fundamentally deceptive', as 'it was not possible to get a real "Balance of Energy". There were tremendous sources of energy which could not be included.'[26] The balance would indeed always be limited to a tiny fraction of global energy flows – namely, technically and commercially produced energy. Eventually, the issue was dropped in 1935, allowing for 'the matter [to] be raised again at a later date'.[27] Only a decade later, it would indeed be on the agenda again.

Powering Economic Reconstruction

The period after World War II is not typically remembered as a beginning of energy governance. The emergence of energy as an object of politics, as well as management of resources with regard to their interrelatedness and common function to perform work, is usually associated with the response to the oil crisis. Timothy Mitchell argues that 'the problem of

25 Haidegger admitted that water power had to be treated differently, as its efficiency was not directly comparable to the efficiency of fuel combustion (he proposed an average annual efficiency factor of 25 per cent for water power); ibid., 28.

26 World Power Conference, 'Minutes of the Meeting of the International Executive Council, Held at the Royal Netherlands Institute of Engineers, The Hague, Holland, on Wednesday, July 17th', 1935, 24.

27 Ibid., 25.

energy as an interconnected and vulnerable system' became a matter of concern in the US in the early 1970s.[28] However, postwar reconstruction posed its own problem of organizing resources for economic growth that led to focus on energy politics and a common treatment of coal, oil, and, later, electricity.[29] This is especially true for Europe, where the coal and power industries lay in shambles after the war. With sky-high government debts, European countries relied on US capital to build up their industrial capacity.[30] Postwar institutions such as the Organisation for European Economic Co-operation and the Coal and Steel Community paid from the beginning attention to a common organization of resources. While access to resources varied, no European state was exempt from solving the problem of powering economic reconstruction. In doing so, they institutionalized ways to deal with the coal, oil, and electricity industries in a collective way.

A view on Haidegger's balance shows that European countries relied on each other's resources. The Ruhr was the largest continental coal-mining area and a source of fuel on which all neighbouring countries depended. For this reason, the destruction of Germany's industrial base as envisioned by the Morgenthau Plan would have affected Europe as a whole (this was one reason why France never supported German deindustrialization).[31] The plan had always been contested, but, when the containment of the Soviet Union began to trump that of Germany,

28 Timothy Mitchell, *Carbon Democracy: Political Power in the Age of Oil* (Verso, 2011), 176.

29 While scarcity was more pressing in Europe, resource surveys were conducted in the US as well; Rüdiger Graf, *Oil and Sovereignty: Petro-Knowledge and Energy Policy in the United States and Western Europe in the 1970s* (Berghahn Books, 2018); Stephen G. Gross, *Energy and Power: Germany in the Age of Oil, Atoms, and Climate Change* (Oxford University Press, 2023); Daniel A. Barber, *A House in the Sun: Modern Architecture and Solar Energy in the Cold War* (Oxford University Press, 2016), 75–82; Richard Lane, 'The American Anthropocene: Economic Scarcity and Growth During the Great Acceleration', *Geoforum* 99 (1 February 2019): 11–21.

30 Ethan B. Kapstein, *The Insecure Alliance: Energy Crises and Western Politics Since 1944* (Oxford University Press, 1990), 48.

31 John Gillingham, *Coal, Steel, and the Rebirth of Europe, 1945–1955: The Germans and French from Ruhr Conflict to Economic Community* (Cambridge University Press, 2004), 101, 153.

US policy turned to focus on aid instead of punishment. The harsh winter of 1946/7 likewise strengthened the belief that Europe needed more long-term US economic assistance if revolt was to be avoided.[32] The recovery of the Western German economy was to facilitate general European reconstruction.[33] The Ruhr was located in the British Sector, but Britain struggled to bring its coal production back to pre-war levels. By 1947, the coal industry was managed in a 'bi-zonal' arrangement with the US.[34] Even before the official start of the European Recovery Program (ERP) in 1948, US capital flowed into European energy infrastructure, including German coal production.[35] While the ERP funds were managed by the Organisation of European Economic Co-operation (OEEC), a French initiative with US backing transformed the Ruhr into a site of European economic cooperation: the European Coal and Steel Community.

France had long debated how it could limit the threat of a neighbour who had attacked it twice within a generation. Access and some degree of control over German coal production were considered crucial to hinder rearmament and secure the French steel industry. Pierre Pflimlin, a French politician, had 'recommended setting up an "energy complex", which would include France and the Benelux nations as well as Germany and be run by a European authority'.[36] In the negotiations with Great Britain and the US, Jean Monnet put forward a proposal to put all management decisions of the German coal industry under an international board, including the US.[37] However, the International Authority of the Ruhr that came to be only specified the amount of coal for export and ensured equal access at stable prices. The Schuman Declaration in 1950 ended a decade of struggle over the German coal

32 Ibid., 116.

33 Ibid., 160.

34 Ibid., 127–8.

35 Ibid., 186; to regain self-sufficiency, especially towards socialist coal from Poland, roughly $400 million of Marshall Plan money flowed into coal and machinery; see Kapstein, *The Insecure Alliance*, 47–9.

36 Cited in Gillingham, *Coal, Steel, and the Rebirth of Europe*, 160.

37 Ibid., 154.

and steel industry. After years of slow rapprochement between French and German industrial and political leaders, France, the Benelux countries, Italy, and the newly formed German government entered negotiations and signed the Treaty of Paris in 1951. The European Coal and Steel Community (ECSC) extended and deepened coal market integration among six European countries over the following years.[38]

Industrialization had been based on coal in most European countries. At the end of the war, they were still largely using coal as a fuel. However, it was a widely shared belief that to regain competitiveness with the US, Europe would need oil – mainly to increase levels of motorization. With Great Britain and France losing influence in the Middle East, Western Europe had no substantive oil production of its own (North Sea oil entered the market only in 1971). Founded already in 1948, the OEEC's Oil Committee was to distribute ERP money to increase European refining capacity. American aid thus actively facilitated Europe's shift to oil while securing a market for its oil companies' new Middle Eastern concessions. Oil was the single largest industry to benefit from ERP money, receiving a tenth of total aid.[39] Though Europe would thus become dependent on a politically volatile region for its fuel, American reserves could act as a strategic stockpile during an emergency.[40]

A peculiar feature of these postwar organizations was their combination of data collection with security mechanisms. Between former enemies, allocation could not be left to politics but had to follow an automatic mechanism based on national data on supply and demand. For the mechanism to work, there had to be clear-cut definitions and production and consumption statistics based on the same coal classification from hard coal to lignite. The ECSC installed expert commissions to gradually harmonize the terms and to standardize resource statistics and coal testing between the six member countries. While the OEEC had no power of allocation, it also had

38 Kapstein, *The Insecure Alliance*, 47.

39 David S. Painter, 'Oil and the Marshall Plan', *Business History Review* 58, no. 3 (1984): 362; Mitchell, *Carbon Democracy*, 29–31; while the American oil industry benefited, European countries pushed for the shift to oil more than the Americans; Kapstein, *The Insecure Alliance*, 66.

40 Kapstein, *The Insecure Alliance*, 69.

permanent commissions on coal, gas, oil, and electricity, which published triannual, harmonized statistics for all receiving countries.[41] When Western oil-consuming countries later institutionalized an emergency sharing mechanism within the OECD, the ECSC's method of resource allocation served as a model.[42] The OEEC's Coal Committee began forecasting Europe's coal production and consumption in 1956; the Oil Committee even began to engage in projections of oil supply and energy demand only five years later, after the Suez Crisis. Both were based on standardized questionnaires sent to the member countries.[43]

But interest in these figures was not confined to the European governments that were struggling to rebuild their industries. At a forum like the World Power Conference, the engineers in public offices at the ECSC, OEEC, or United Nations Economic Council for Europe came together with those working in industry or state-planning agencies in the East. In a paper on 'World Energy Statistics', G. H. Daniel, from the British Ministry for Fuel and Power, explained the general interest in statistical information on energy:

> The oil company considering its likely markets, the bank scrutinising the case for a hydro-electric scheme, the aluminium company looking for a location for a large smelter, the engineering firm considering the likely market for oil-fuelled boilers, and the Government concerned about the industrial prosperity of its country are all interested in the

41 Jan van den Heuvel, 'Contribution to Discussion on "The Methods of Compilation and Use of Statistics in the Production and Utilization of Energy"', in World Power Conference, *Fünfte Weltkraftkonferenz Wien*, vol. 4, *Statistische Methoden in der Energiewirtschaft* [Fifth World Energy Conference, vol. 4, The Methods of Compilation and Use of Statistics in the Production and Utilization of Energy] (Österreichisches Nationalkommitee der Weltkraftkonferenz, 1957), 1099–100; Hans Freytag, 'Contribution to Discussion on "The Methods of Compilation and Use of Statistics"', in World Power Conference, *Fünfte Weltkraftkonferenz Wien*, vol. 4, 1097–9.

42 Graf, *Oil and Sovereignty*, 51.

43 Henning Türk, 'The Oil Crisis of 1973 as a Challenge to Multilateral Energy Cooperation Among Western Industrialized Countries', *Historical Social Research* 39, no. 4 (2014): 211–12.

> same basic material – data on energy resources, production, consumption, and price.[44]

These statistics had to go beyond single resources, as industries were becoming increasingly interdependent. Seen from a technical point of view, there were now several options available for almost all uses of nature's work. The US market had long seen competition between domestic coal and oil. In Europe, which had no sizeable oil production of its own, there was at first no direct competition between imported oil and domestic coal industries. Oil was refined into gasoline to power cars, trucks, tractors, and planes, while coal was used for steelmaking, for heating, and in chemical industries. However, the production of gasoline yields a by-product that can be used as fuel oil. Over the 1950s, fuel oil slowly entered the European heating market. While some countries shielded their domestic coal or coal-trading industries by taxing fuel oil, others did not.[45] With ongoing electrification and nuclear power on the horizon, engineers, managers, and planners registered that the space for 'interchangeability' or 'substitutability' was growing. This meant that it was no longer possible to treat the coal, oil, and electricity industries as independent from each other: Whoever invested in one of them had to take into account developments in the others. A US study from 1950 estimated that two-thirds of energy commodities and 90 per cent of primary sources were substitutable – that is, were subject to inter-energy market competition. Only three areas of relative noncompetition remained – electricity, gasoline (in cars and aircrafts), and coke for blast furnaces – but who knew for how long?[46]

By the 1950s, the European energy problem had shifted from reviving and organizing the coal and steel industries to marshalling energy for

44 G. H. Daniel, 'World Energy Statistics', in World Power Conference, ed., *The Transactions of the Fifth World Power Conference*, vol. 4 (Österreichisches Nationalkommitee der Weltkraftkonferenz, 1957), 958.

45 Gross, *Energy and Power*, 33; N. J. D. Lucas, *Energy in France: Planning, Politics and Policy* (Europa Publications, 1979).

46 Harold J. Barnett and United States Bureau of Mines, *Energy Uses and Supplies, 1939, 1947, 1965* (US Governmental Printing Office, 1950), 7–9.

economic growth. After the Korean War had sent oil prices upwards, the OEEC began first to talk of a 'European energy problem'. Louis Armand, a French engineer of railway electrification who would move on to become head of Euratom, was commissioned to draft a first outline of the problem in 1955. In the report, he described Europe's position at a 'turning point', about to fully enter the electric and petroleum age, but held back by comparatively high costs of fuel and power.[47] Armand recommended the appointment of a group of experts to study the energy problem more deeply, especially regarding nuclear power as a solution. This commission, headed by the British chemist Harold Hartley, produced the first pan-European energy forecast in 1956, published as *Europe's Growing Energy Needs: How Can They Be Met?* Estimating the primary energy production and needs for each country separately, the study then aggregated them to gain a 'reasonable' picture 'of the energy economy of Western Europe as a whole' for the years 1960 and 1975.[48] To add up the statistics of coal, gas, oil, and hydropower, the report converted all primary energies into 'tons of coal equivalent' (toc) – a unit that had been defined in the OEEC. By doing so, it not only published the first pan-European energy balance but also made it possible to envision a future where forms of energy would be combined differently to satisfy a growing European energy demand.

The electric and nuclear industries were at the forefront of this statistical endeavour. As it had to account for grids that integrated different power sources, the electric industry had long developed methods of economic comparison between electrical energy hydraulically and thermally produced.[49] The Union Internationale des Producteurs et

47 Organisation of European Economic Co-operation, *Some Aspects of the European Energy Problem: Suggestions for Collective Action Report Prepared for the O.E.E.C. by Mr. Louis Armand, Chairman of the Board of the Société Nationale des Chemins de Fer Français* (Organisation for European Economic Co-operation [OEEC], 1955).

48 Organisation of European Economic Co-operation and Harold Hartley, *Europe's Growing Needs of Energy: How Can They Be Met?* (Organisation for European Economic Co-operation [OEEC], 1956), 21.

49 World Power Conference, *Transactions of the World Power Conference, Sectional Meeting, Basle 1926*, vol. 2, sec. C, *The Economic Relation Between Electrical*

Distributeurs d'Energie Electrique, a Europe-wide association of the electric industry, had long worked to standardize those methods. Nuclear power was more than an additional source to account for: The volume and duration of investment into nuclear power required state-level energy forecasting.[50] Nuclear power promised a partial independence to European countries, or at least a distribution of dependencies. While it could not entirely make up for oil (not in transport and heating), a reliable source of domestic electricity, undisrupted by the strikes haunting the coal industry, could still contribute significantly to a state's infrastructural power. The size, costs, and risk of nuclear power required not only the involvement of the state but also some degree of supply-and-demand planning. It implied a certain nexus between politics, science, and energy, encompassing methods of electricity forecasting, research institutes, demonstration reactors, and administrative bodies overseeing the industry. Many of the engineers who developed the balancing method for forecasting were related to these institutes.[51]

Energy Produced Hydraulically and Electrical Energy Produced Thermally (Birkhäuser & Cie, 1927).

50 For this new kind of 'nuclear futurology', see Jean-Baptiste Fressoz, *More and More and More: An All-Consuming History of Energy* (Penguin, 2024), 142–59.

51 Sheila Jasanoff and Sang-Hyun Kim, 'Containing the Atom: Sociotechnical Imaginaries and Nuclear Power in the United States and South Korea', *Minerva* 47, no. 2 (26 June 2009): 119–46; Frank Uekötter, *Atomare Demokratie: Eine Geschichte der Kernenergie in Deutschland* (Franz Steiner Verlag, 2022). In France, the Commission d'Energie Atomique was mandated to develop the nation's energy system; Gabrielle Hecht, *The Radiance of France: Nuclear Power and National Identity After World War II* (MIT Press, 1998), 27. Energy balancing was part of French planning work at least since the 1960s; see Lucas, *Energy in France*, 111. In Germany, the industry itself – particularly the gas and electric companies who were vulnerable to competition from other sources – set up research institutes as early as the 1940s, focusing on relations between energies and the study of consumption. But it was the nuclear power research institute (Kernforschungsanlage Jülich, KFA) which conducted pioneering research into the energy systems models in the 1970s (the model MARKAL [Market Allocation] was developed in a cooperation between KFA and the Brookhaven National Laboratory in the US).

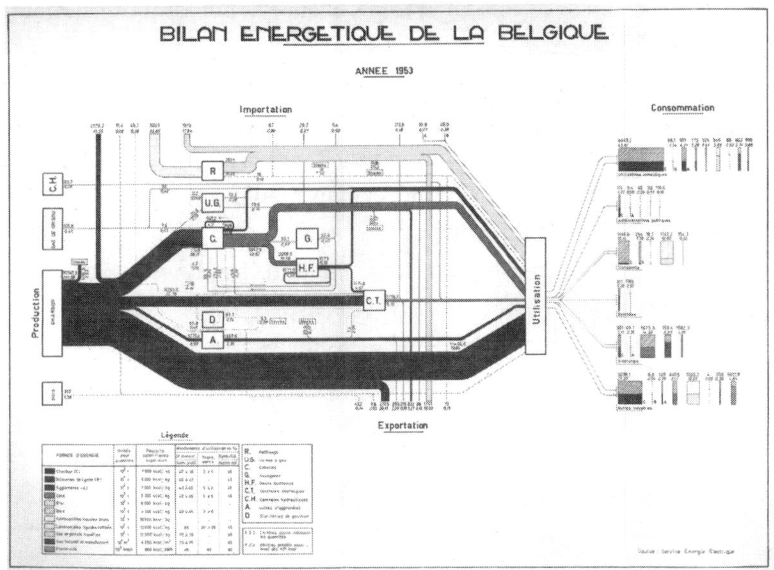

Figure 5.2. Belgian energy balance for the year 1953. Balances over the years show how refinery capacity and oil consumption increased in the 1950s.

How to Treat Energy Collectively

Engineers in public office and private industry agreed that security of energy supply in times of economic growth could no longer be ensured by looking only at a single resource. The balance method promised a solution to this problem. By compiling a common balance for all resources performing work, those resources became appearances of the same value – energy. Balances thus went beyond the gathering of individual resource statistics and their aggregation into world energy totals: They would capture the coal, oil, gas, and electricity industries in their interconnection and in their respective relation to the forms in which nature's work was used. Over the 1950s and '60s, corporations, regulatory agencies, and government bodies began to turn to energy balancing, only to find out how difficult it was to treat energy collectively.

Everything could be expressed in terms of energy, but for those numbers to be economically meaningful, decisions had to be taken on the limits of the energy economy and the valorization of its forms.

In 1950, the young economist Harold J. Barnett had modelled the US energy economy for the Department of the Interior. For this report, Barnett had used the computer-based input-output model developed by Wassily Leontief, his PhD supervisor at Harvard, to formalize a functional understanding of resources that he had picked up from German American economist Erich Zimmermann.[52] Widely influential, *Energy Uses and Supplies (1939, 1947, 1965)* modelled the energy sector as a modular system where single parts could be replaced with others. Barnett distinguished between primary sources (coal, crude oil, water power) and energy commodities (electricity, diesel, coke, etc.), treating the latter as both outputs of primary sources and inputs into other sectors. 'The economy's functional requirements for energies could be satisfied with a variety of energy commodity combinations', Barnett wrote. In turn, 'these commodities could themselves be produced from a variety of domestic natural resources and imports. Therefore, as a practical matter, projections must be made for all the energy supplies simultaneously, as each projection depends on the others.'[53] This diagnosis was shared by the speakers on 'Statistical Methods in the Energy Economy' at the World Power Conference's Vienna congress (Harold Hartley, head of the OEEC's energy commission, was by then president of the WPC).[54] If demand for one energy product depended on another, forecasting required a model that could account for the 'complex' these energies formed. 'Seen from the statistical point of view', summarized an Austrian mathematician, 'the most comprehensive foundation for advance planning of the energy economy is the input-output-analysis.'[55]

52 Barber, *A House in the Sun*, 79–80.

53 Barnett and United States Bureau of Mines, *Energy Uses and Supplies*, 7.

54 Note that at that time the German original uses the term 'energy economy' (*Energiewirtschaft*), which was common since the 1930s, whereas the English translation rendered it 'The Methods of Compilation and Use of Statistics in the Production and Utilization of Energy'.

55 Leopold Schmetterer, 'General Report on "The Methods of Compilation and Use of Statistics in the Production and Utilization of Energy"', in World Power Conference, *Fünfte Weltkraftkonferenz Wien*, vol. 4, 798.

While the industrialized and industrializing world now agreed on the need for energy balancing, the statistical basis for them was still missing, as few countries could furnish production and consumption statistics across all forms of energy. Nathaniel B. Guyol, a Canadian geographer who had moved from the Tennessee Valley Authority to the UN Statistical Department, gave a passionate presentation on 'Current Problems in the Compilation of World Energy Statistics'. A supporter of an energy-based measure of economic development, Guyol was responsible for setting up a statistical apparatus on energy data.[56] Surveying the statistical efforts of the World Power Conference, the Union Internationale des Producteurs et Distributeurs d'Énergie Électrique (UNIPEDE), and others, he concluded that data on production was relatively comprehensive, while data on consumption was sparse. Only coal and electricity statistics were almost complete. Guyol admitted that at this stage, a large part of the globally expended energy could not be traced at all.[57] Yet without proper production and consumption statistics, which included data from large-scale industry based on measurements, energy balances would not be very meaningful.

Beyond the availability of statistics, energy balancing also raised tricky methodological questions with political implications: What constituted an energy use? Which categories of use should be distinguished – if one kept in mind that they also had to be measurable?

56 Thomas Turnbull, 'From Incommensurability to Ubiquity: An Energy History of Geographic Thought', *Journal of Historical Geography* 73 (July 2021): 18–19; Alan Clarke and Judy Trinnaman, 'Developing Comprehensive Energy Balances', *Energy Exploration and Exploitation* 5, no. 3 (June 1987): 175–85; Nathaniel B. Guyol, *Energy Interrelationships: A Handbook of Tables and Conversion Factors for Combining and Comparing International Energy Data* (National Energy Information Center, 1977).

57 Nathaniel B. Guyol, 'Current Problems in the Compilation of World Energy Statistics', in World Power Conference, *Fünfte Weltkraftkonferenz Wien*, vol. 4, 972. The UN General Assembly would only in 1976 agree 'that the development of a system of integrated energy statistics should have high priority in the Commission's programme of work' and that 'energy balances [are] the key instrument in the coordination of work on energy statistics and the provision of data in a suitable form for understanding and analysing the role of energy in the economy'; United Nations Statistical Division, *International Recommendations for Energy Statistics*, 2.

Everybody could agree that it should encompass the application of mechanical work and heat, but other uses were more contentious. In his work on a German energy balance at the Technical University Karlsruhe, Herbert Müller argued for the inclusion of sound and ultrasound – essentially making the media and communication industry part of the energy economy (had this classification prevailed, it would be possible to read the energy consumption of the digital economy from a balance sheet today). At the same time, he was reluctant to include a category of chemical energy, fearing that the balance would then have to cover the myriad of different reactions taking place in chemical industry. Wilhelm Frank, one of the few communist officials in Austria's Ministry for Trade and Reconstruction, pointed out that those decisions were essentially political: They depended on the differences a state deemed necessary to govern the energy economy. In line with the Marxist tradition of energy balancing, he suggested that this could include the distinction between a reproductive and a productive use of energy – something the Soviet statisticians of the 1920s had hoped to achieve as well.[58]

Most vexing was the problem of energetic valuation. The principle of energy conservation proclaimed an equity that belied the technical side of things. The chaotic diversity of units was only the superficial expression of this, clouding the view for the deeper and more serious difficulties. As one of the founders of Électricité de France, Pierre Ailleret, put it: Anybody who deals with problems of the energy industry recognizes their complex, entangled nature and will seek to overcome them by searching for a common standard of valuation. 'It would be tempting to pretend that all these forms of energy have the same dimension and to look for equivalent values.' Values that, though depending on the units chosen, ultimately remain the same over time and space. 'Unfortunately, the deeper reality does not conform to this rather simple scheme.'[59] The crux of balancing was, in Barnett's words, 'the discrepancy between total energy supply used

58 Wilhelm Frank, 'Strukturbilanzen der Energiewirtschaft', *Unternehmensforschung* 2, no. 3 (1 September 1958): 125–6.

59 Pierre Ailleret, 'Bewertung verschiedener Energieformen', *Archiv für Energiewirtschaft* 13, no. 2 (1959): 44.

and total useful work performed'.[60] The energy content is no natural constant and the energy derived as useful work is conditioned not only by physical but also by technical, economic, and social conditions.

One way to capture those was the use of *technical efficiencies* rather than physical energy values. When the Hartley report expressed water power in the unit of Europe's major source of energy, tons of coal equivalent, it made use of a technical conversion factor. Water power did not undergo the same process that coal did: It drove a turbine, whose movement powered a generator, which set off an electric current. Nothing was combusted, and – apart from friction – there was no heat involved that could have been measured and converted into coal equivalents. Instead, coal equivalents were defined in the report as 'the amount of coal required to produce the amount of energy furnished by other sources'.[61] The production of a unit of electricity required more coal than water power in terms of its physical energy content. Put differently, hydropower was able to replace more coal than its physical energy value indicated – just how much depended on the efficiency of thermal plants. This problem of equivalents affected the entire balance. Barnett gave the example of refined fuels, where '1 B.t.u. of Diesel oil performed as much work as 5 B.t.u. of fuel oil in 1947'.[62] But only in the US, and only in 1947! If efficiencies determined the relative size of energies to each other, and those efficiencies changed over time and space, then energy balances would be incomparable, energetic value in constant flux. Equivalence across time and space would be impossible to achieve.[63]

Taken seriously, the question of efficiency could easily lead down a rabbit hole. Herbert Müller gave a passionate speech against what he saw as a sloppy use of energy balancing at the WPC's Vienna conference in 1956. As head of the Energieforschungsstelle at the Technical University in Karlsruhe, he had been in contact with Wassily Leontief, Harold Barnett, and Erich Zimmermann. Frustrated by how everyone was

60 Barnett und United States Bureau of Mines, *Energy Uses and Supplies*, 13.

61 Organisation of European Economic Co-operation and Hartley, *Europe's Growing Needs of Energy*, 17.

62 Barnett and United States Bureau of Mines, *Energy Uses and Supplies*, 4.

63 Ailleret, 'Bewertung verschiedener Energieformen', 44.

calling for energy balances as if statistical comparability between energies had already been achieved, he fired a salvo of methodological questions at the audience. In his view, the input-output table could not lead to useful results under given conceptual and statistical conditions. 'It is impossible to measure and value different units of energy in a definite technical manner', he exclaimed. 'It is even less possible to measure and value them in a definite economic manner, and it is utterly impossible to derive exact forecasts from what is measured.'[64] Technical efficiencies, like the ones estimated in the Barnett and Hartley reports, were not even the real problem, as most losses happened in operation. If energy balances were to be meaningful at all, they had to be anchored in actual, repeated measurements of efficiency and useful work expended in production. 'We must go into the factories . . . Only by analysing useful energy can we achieve a clearer balance!'[65]

Wilhelm Frank put it similarly: A truly critical observer could not be happy with energy balances based on technical efficiencies.[66] On the one hand, when hydropower was valued according to the amount of coal it could replace, instead of representing the physical process that happened when water power was turned into electricity, it was overwritten with a heat process that did not actually take place, smuggling losses into the balance that did not exist. On the other hand, when electricity and fuels were assigned values according to their physical energy content, they were compared only in terms of quantity of energy not thermodynamic possibilities.[67] Drawing on the concept of a 'maximum theoretical useful work', the Austrian Federal Ministry for Trade and Reconstruction compiled an exergy balance alongside the energy balance: a balance that related each conversion process to its thermodynamic optimum.[68] As a

64 Herbert Müller, 'Contribution to Discussion on "The Methods of Compilation and Use of Statistics in the Production and Utilization of Energy"', in World Power Conference, *Fünfte Weltkraftkonferenz Wien*, vol. 4, 1107.

65 Ibid., 1007–8.

66 Frank, 'Strukturbilanzen der Energiewirtschaft', 129.

67 Ailleret, 'Bewertung verschiedener Energieformen', 44–5.

68 These questions had long been discussed in thermodynamics and thermal engineering under the concept of 'work capacity' or 'useful energy'. Since the 1930s, a lively research field on 'technical work capacity' (*technische Arbeitsfähigkeit*) had

result, overall efficiency declined from 41 per cent in the energy balance to only 26 per cent in the exergy balance.[69]

The point of energy balancing was not to arrive at an *energetically* optimized economy. Rather, the costs of investment into energy made it necessary to step up material accounting: Energy balances could show whether it was worth it to get rid of losses or switch to a different fuel. This became particularly evident in the way in which entropy balances were discussed. Taking entropy seriously did not mean posing limits to economic growth (as is today sometimes suggested). Quite the opposite: An exergy balance was an attempt to approximate the energy usefully employed to arrive at a better estimation of real energy demand. As Pierre Ailleret pointed out:

> We do not use thermodynamically perfect machines, but boilers and turbines that are anything but perfect, and do not even strive to be, because they are geared towards economic optima, which are not just about efficiency, but also about investment costs.[70]

For the UNIPEDE, Ailleret had developed a method to calculate the economic value of a calorie as a function of its temperature. By doing so, he could show that each unit of energy was not the same in Europe: Calories differed in their economic value. A thermodynamically optimal economy was neither realistic nor desirable, but it was useful as an outer border of what was possible.

Forecasting threw up an even deeper problem of valuation: What counts as energy in the future? Only those resources that were used now

---

formed in the Central European countries, driven mostly by the Croatian Franjo Bošnjakovic and the Slovenian Zoran Rant (who coined the term 'exergy'). These concepts referred to the 'maximum theoretical useful work obtained' during energy conversion – i.e., when a system is brought in the thermodynamic equilibrium. Göran Wall, 'A Brief Commented History of Exergy: From the Beginning to 2004', *International Journal of Thermodynamics* 10, no. 1 (2007): 3–5.

69 Wilhelm Frank, 'Contribution to Discussion on "The Methods of Compilation and Use of Statistics in the Production and Utilization of Energy"', in World Power Conference, *Fünfte Weltkraftkonferenz Wien*, vol. 4, 1108–9.

70 Ailleret, 'Bewertung verschiedener Energieformen', 45.

or those that could possibly be used? What about resources that could perform work in more than one way? Wilhelm Frank, highlighting the many hidden decisions inherent in compiling a balance, gave the example of natural gas. The straightforward way was to account for it in terms of its volume and heating value. But the heating value is no natural constant; its measurement varies with the environment (see chapter 3). A very critical energy engineer might pose even more heretic questions when thinking of the farther future: Should not natural gas also be valued as a potential source of heavy hydrogen, an input in fusion reactors? Why not add this fusion energy to that of natural gas as a precautionary measure?[71] Frank's example, though bordering on the outlandish in 1958, pointed to a real experience of mutability of resources at the atomic level at the time. As Harold Barnett and Chandler Morse pointed out in their 1963 study *Scarcity and Growth*, stones could transform into springs of energy: 'Two decades ago Vermont granite was only building and tombstone material; now it is a potential fuel, each ton of which has a usable energy content (uranium) equal to 150 tons of coal.'[72]

These debates on the basic compilation principles of energy balances were not decided in the 1950s, and engineers even valued this methodological diversity. Regardless of the concrete methodology chosen, however, the aim was to understand, and govern, how forms of energy could be replaced and combined differently. Energy balancing responded to the problem and possibility that came with an increased space of substitutability between forms of energy. In compiling a balance, the current and future energetic value of resources is determined in a specific way. As should be clear from the debates of the 1950s, balances do not *represent* a self-evident unit. Far from a pure scientific process approximating its object, balancing work required decisions on what should be part of this statistical snapshot of the energy economy and how it should be valued. When balances were used to forecast, it also implied decisions on how the uncertainty of the future should be integrated.

71 Frank, 'Strukturbilanzen der Energiewirtschaft', 127.

72 Harold J. Barnett and Chandler Morse, *Scarcity and Growth: The Economics of Natural Resource Availability* (RFF Press, 1963), 7.

Engineers and economists abstracted a model of the 'energy economy' from the messiness of material and energetic relations and transformations according to certain rules. There was no single balance, only balances from different points of view – and this diversity allowed insights that single balances did not.[73] This diversity of methodologies gave way to a formalization of substitutability in the wake of the oil crisis.

Formalizing Substitutability

Over the 1950s and 1960s, economic growth became the common project of the European OEEC and then the transatlantic OECD.[74] Richard Lane argues that resource economics, by shifting to a functional understanding of resources and a price-based definition of scarcity after World War II, enabled a dematerialized understanding of economic growth that justified the growth paradigm.[75] As can be seen from the discussions among engineers presented above, they too relied on a functional understanding of energy in a double sense: First, that the availability and pattern of use of primary and secondary energies would be determined by what was technically possible, and, above all, by what was economically rational. And, second, that society needs no longer one particular resource (oil, gas, coal, or electricity) but a service (heat, light, traction, etc.) that could be performed by a range of resources. Speaking of energy at all, rather than concrete resources, already entails this functional understanding. 'Nature's input should now be conceived as units of mass and energy, not acres and tons', wrote Barnett and Morse.[76]

73 Ailleret, 'Bewertung verschiedener Energieformen', 47.

74 Matthias Schmelzer, 'The Growth Paradigm: History, Hegemony, and the Contested Making of Economic Growthmanship', *Ecological Economics* 118 (October 2015): 262–71; Matthias Schmelzer, *The Hegemony of Growth: The OECD and the Making of the Economic Growth Paradigm* (Cambridge University Press, 2016).

75 Lane, 'The American Anthropocene'.

76 Barnett and Morse, *Scarcity and Growth*, 238. This molecular view of nature as recombinable was later taken even further by Julian Simon, for whom the only resources needed were human genius, matter, and energy – the 'master resource' that could melt nature into entirely new forms. Troy Vettese, 'The Master Resource:

While we can observe the reduction of the conceptualization of a natural moment in economics – especially after 'growth' becomes an explicit issue – such ideas were not free-floating.[77] With regard to energy, the abstraction was rooted in the substitutions that took place, and that threw up real social and economic problems, just to be reified into an abstract substitutability when extended in time and space.

Energy balances were an attempt to come to terms with the increasing fungibility of energy commodities. They reflected the insecurity of investment at a time when different forms of energy became more widely available. After the entry of oil and the integration of the coal market, the European community pushed for deeper integration of other energy markets in the 1960s. They made plans for natural gas from the Soviet Union; nuclear power was on the horizon.[78] A similar shift from coal to oil and gas took place in the East as well.[79] Industries, whether nationally or privately owned, could no longer take decisions without taking various forms of energy into account: The electricity corporation pondering a hydroelectric scheme was looking for the sweet spot where the replacement

---

Energy, Inter-Planetary Capitalism, and Neoliberal Cornucopianism', in Daniela Russ and Thomas Turnbull, eds, *Energy's History: Toward a Global Canon* (Stanford University Press, 2025), 202–21.

77 Both sides of the debate on economic growth that ensued over the 1970s relied on an abstract understanding of energy. The common treatment of coal, gas, oil, and electricity was rooted in substitutions that actually took place, but, as we have seen, energy forecasts went beyond what was technically possible and economically feasible at any given time. The Club of Rome scientists sought to challenge this reliance on future technological change, so the Limits of Growth model did not account for substitutability at all. Energy was treated as too abstract for the model to be considered an energy model by later systems scientists. At the same time, the systems scientists of the 1970s were no less susceptible to abstractions: Unlike the engineers of the 1950s, who debated whether and how future technologies should be factored into the valuation of each resource, they hypostasized the mutability of resources into an abstract process of technical change. Christophe Cassen and Béatrice Cointe, 'From the Limits to Growth to Greenhouse Gas Emissions Pathways: Technological Change in Global Computer Models (1972–2007)', *Contemporary European History* 31, no. 4 (November 2022): 614.

78 Per Högselius, *Red Gas: Russia and the Origins of European Energy Dependence* (Palgrave Macmillan, 2013), 3.

79 Jeronim Perović, ed., *Cold War Energy: A Transnational History of Soviet Oil and Gas* (Palgrave Macmillan, 2017).

of the thermal calories by hydroelectricity was economically optimal; the oil industry could glimpse from energy balances the size of a country's heating or transport market not yet served by oil; and the coal industry could use balances to convey their vital importance to the nation's economy to politicians. Countries had become energy suppliers and markets for each other and acted like it. Without the broad interest in these numbers, and their importance for navigating a politics of growth under heightened competition, it is doubtful so much intellectual work would have been expended in compiling them. Energy balances reflect this abstract view of industrial capital, for which energy had become a factor of production among others. The ultimate horizon was a global energy balance, where a calorie of Indonesian coal and a kilowatt-hour of Norwegian water power could be related in terms of their potential of 'useful work'.

The form a world energy balance would take was still open, but it was already a world of energy balances in the 1960s. While methodologies differed, it had become standard for all self-respecting industrialized and industrializing countries to account for their energies. 'In every country the framework of energy evaluation is the energy balance sheet', as a Czech delegate to the World Power Conference in 1962 put it matter-of-factly.[80] The method was shared as a statistical tool to govern industrialization across the West, East, and South. For the socialist countries, undergoing their own shift to gas and oil, energy balancing was part of economic planning, a method through which different patterns and proportions of sectors could be envisioned. In theory, the optimal solution would then guide planning.[81] For developing countries, it served to direct and evaluate industrialization – understood as the replacement of 'older' (coal, wood) for 'newer' (electricity) forms of energy. In the West, it provided market

80 Florian Kucera, 'Economic Evaluation of Energy', in World Power Conference, ed., *Transactions of the Sixth World Power Conference, Melbourne, Australia* (Australian National Committee of the World Power Conference, 1962), 4207. The EC began publishing a European energy balance that harmonized existing national balances in 1965.

81 Lev A. Melentev, 'General Report: Energy Balances', in *Transactions of the Seventh World Energy Conference [Sedmaia Mirovaia Energeticheskaia Konferenciia]*, sec. B, *Energy Balances* (Sovetskii Nacinonalnii Komitet Mirovoi Energeticheskoi Konferencii, 1969), 46.

information for managers to navigate competition and for politicians to ensure an 'orderly progress of substitution'.[82] But there was no simple way to integrate national energy balances: Meaningful balances depended on a carefully chosen method of evaluation, evaluation in turn hinged on efficiencies, and efficiencies differed widely across countries depending on the structure, age, and handling of energy technology. There was a general agreement that the 'total amount of energy usefully employed' was not going to be internationally comparable in the foreseeable future, as it would require standardized, regular measurements at the level of factory floors. Whether without it balances would be meaningful and comparable between countries was disputed, but the oil crisis made these doubts look small-minded and misplaced: In the 1970s, the diversity of balancing on the national level gave way to standardization beyond the OEEC countries through the dominance of one particular perspective – oil substitution.

A potential disruption of oil supply had long been anticipated, when the Arab members of OPEC boycotted Western supporters of Israel in the Yom Kippur War in 1973. The threat of oil shortages and rising prices had been a recurring issue in the OECD's Oil Committee. The pre-1970s emergency mechanism (restricted to the Western European countries) was based on national strategic stockpiles, which could be shared once a crisis had been declared. Officials of multilateral and national institutions expected and prepared for an oil shortage, and energy system analysts had simulated all kinds of changes in national energy balances. As we have seen, the shock did not create the energetic perspective as such – the collective governance of resources that perform work had already been ongoing. However, it accelerated the integration of several policy fields into 'energy policy' that had already started before, and it prompted international standardization, first among OECD countries and then beyond. It catapulted the 'energy problem' onto the agendas of existing international organizations and sparked new ones.

82 S. Balke, 'Problems of Coordination and Substitution in the Production and Utilization of Energy', in World Power Conference, *Transactions of the Sixth World Power Conference, Melbourne, Australia*, 4182; William A. Vogely, 'Analytical Uses of Energy Balances', in *Transactions of the Seventh World Energy Conference*, sec. B, 185–97.

The 1973 oil crisis should be understood as only one moment in a longer conflict about resource sovereignty.[83] Faced with a changing power in the Middle East and a loss of control over the flow of oil, the US pushed for a deeper consumer integration and a restructuring of national energy policies along similar lines. Formulated at the Washington Energy Conference in 1974, this new programme of consumer cooperation became known as the International Energy Program (IEP). An autonomous body within the OECD, the IEA was to administer the programme. Continuing the OECD's soft governance by numbers, the IEA put forth global energy balances as an instrument to regain the political capacity to act in the face of a confusing interdependency of global resource flows. Along with larger stockpiles and an allocation mechanism, the new agreement brought in measures such as energy conservation, the development of alternative energy sources, and the acceleration of energy research and development, to reduce dependence on oil.[84] The IEP drew mainly from earlier schemes of multinational resource allocation, but not only: It also introduced the new concept of 'fuel-switching capacity'. Instead of holding reserves, countries could 'hold' the capacity to quickly substitute one fuel for another in the case of a shortage.[85] The idea was that not only stocks but also the ability to rapidly replace oil with another source of energy would decrease demand for petroleum in case of an emergency. The alternative fuel was to be expressed in terms of oil

83 Christopher R. W. Dietrich, *Oil Revolution: Sovereign Rights and the Economic Culture of Decolonization, 1945 to 1979* (Cambridge University Press, 2017), 264; Giuliano Garavini, 'Completing Decolonization: The 1973 "Oil Shock" and the Struggle for Economic Rights', *International History Review* 33, no. 3 (September 2011): 473–87; Graf, *Oil and Sovereignty*; Jonas Kreienbaum, *Das Öl und der Kampf um eine neue Weltwirtschaftsordnung: Die Bedeutung der Ölkrisen der 1970er Jahre für die Nord-Süd-Beziehungen* (De Gruyter, 2022).

84 Türk, 'The Oil Crisis of 1973', 221.

85 Thijs Van de Graaf and Dries Lesage, 'The International Energy Agency After 35 Years: Reform Needs and Institutional Adaptability', *Review of International Organizations* 4, no. 3 (1 September 2009): 301; J. C. Woodliffe, 'A New Dimension to International Co-operation: The OECD International Energy Agreement', *International and Comparative Law Quarterly* 24, no. 3 (July 1975): 528; Richard Scott and International Energy Agency, *The History of the International Energy Agency, 1974–1994: IEA, the First 20 Years* (OECD/IEA, 1994), 414.

equivalent and would be credited towards the emergency reserve commitment. In practice, fuel-switching capacity accounted only for a minor part of those countries' emergency reserve which used petroleum for power production. Only here, petroleum could be easily replaced (with natural gas, which could easily be expressed in tons of oil equivalent). More importantly, however, the IEP formalized the substitutability between 'energies' for the first time on a multinational level.

As the OECD and many Western European governments saw it, none of their countries could secure their oil supply unilaterally. In this situation, sovereignty became less a matter of direct control of oil, rather than one of knowledge of the oil market in particular and energy markets in general.[86] While the OECD had collected data on energy production and consumption since 1962, it began to harmonize energy balances between member countries in 1976. Unable to standardize the model beyond its member countries, the OECD still tried to choose its conversion factors in a way that they could potentially be internationally harmonized. It did so by reaching out to a couple of international technical organizations, among them the World Energy Conference (WEC), the Union Internationale des Producteurs et Distributeurs d'Énergie Électrique (UNIPEDE), and the International Atomic Energy Agency (IAEA). The prominence of organizations dealing with electricity production is not arbitrary; it speaks to the fact that the electric industry had long practised accounting of power-source substitution. What is more, electricity, particularly produced from nuclear power, was understood as an important 'medium' of fuel substitution.[87] The organizations formed a working group to study conversion factors between fuels and electricity at the production and consumption side. By studying processes where fuels and electricity could directly be compared – the combined production of heat and electricity (CHP) or heat pumps – they hoped to arrive at average conversion factors.

86 Graf, *Oil and Sovereignty*, 286.

87 L. C. Bateman, 'Editorial Review: Conservation of Energy Utilization', in *Energy Resources: Availability and Rational Use, a Digest of the 10th World Energy Conference, Istanbul, 19–23 September 1977* (IPC Science and Technology Press, 1978), 36.

As the result of this working group, a first international agreement on conversion factors was reached in 1985. A little handbook, entitled *Substitutions Between Forms of Energy and How to Deal with Them Statistically*, jointly published by the two organizations in 1985, established a conversion factor that should be used in energy balancing.[88] This field of substitutions is defined as the technical processes where fuel can be replaced: in any thermal processes, heating, or the generation of electricity. Here, the varying efficiencies have to be taken into account, 'as electricity is not used as a fuel; heating can be restricted to the precise spot where it is required, thermal insulation is much easier to provide, there is no heat loss through the evacuation of products of combustion, control is more precise'.[89] For all these reasons, electricity replaces more fuel joules than it adds in electricity – 2.6 times more, according to UNIPEDE and the WEC. As absurd as this might sound, it reflected the inequality of technically mediated energy. While this factor was somewhat arbitrary (and subject to technical change), UNIPEDE and the WEC argued that it would be 'far less in disagreement with commercial values' than the results of other methods.[90]

A World Plan for Energy

The higher level of oil prices after the 1970s had consequences beyond Western oil consumers and Middle Eastern oil producers. While it affected the balance of payments in all consumer countries, poorer countries were hit particularly hard. In the 1970s, various schemes and programmes were developed to deal with the consequences of the oil crisis for 'non-oil-producing developing countries', including the

88 World Energy Conference and International Union of Producers and Distributors of Electrical Energy, *Substitutions Between Forms of Energy and How to Deal with Them Statistically: A Guide* (World Energy Conference in collaboration with the Union Internationale des Producteurs et Distributeurs d'Énergie Électrique, 1985).

89 Ibid., 8.

90 Ibid., 10.

creation of funds to funnel OPEC oil revenues into developing countries.[91] The resolution coming out of the North–South dialogue proposed an international energy institute 'to assist all developing countries in energy resources research and development', a possibility that remained unrealized.[92] As a tool to govern substitution, economic interdependence, and investment, energy balancing meant to play a role in developing countries' response. Unfortunately, the energy data of many countries remained 'inadequate to pinpoint areas where oil substitution could be realized'.[93] It was precisely in this context that energy balances were promoted as a statistical innovation that would help non-OECD countries realize oil substitution and rationalize their energy use.[94] The Organización Latinoamericana de la Energía (OLADE), founded as the Latin American consumer organization along with the IEA in 1973, integrated energy balancing into its programme from the beginning.[95] While OLADE chose a slightly different method to valuate energies, the general framework of the balance sheet was modelled after the OECD balance, which was recommended by UN organizations.[96] Both the UN and the IAEA offered training courses and published handbooks for energy

91 Woodliffe, 'A New Dimension to International Co-operation', 525; David M. Wight, *Oil Money: Middle East Petrodollars and the Transformation of US Empire, 1967–1988* (Cornell University Press, 2021), 85–94; Richard Bailey, 'Development and Energy Costs: A Third World Perspective', *Third World Quarterly* 1, no. 4 (1979): 65–7.

92 Results of the Seventh Special Session of the UN General Assembly, Resolution 3362 (S-VII), adopted 16 September 1975.

93 World Energy Conference, 'Minutes of the International Executive Council, Held at the Deutsches Museum, Munich, Federal Republic of Germany on Saturday, 6th and Sunday, 7th September, 1980 and at the Conference Centre of the Munich Messegelände on Friday, 12th September', 1980, 9.

94 Craig S. Bamberger, *The History of the International Energy Agency: The First 30 Years*, supplement to vols 1–3 (International Energy Agency, 2004), 306–7.

95 Luiz Augusto Marciano da Fonseca, 'National Energy Plans in the Asia-Pacific Region: The Latin American Energy Organization (Olade)', *Energy* 6, no. 8 (1 August 1981): 862.

96 Pierre Vernet, 'Energy Balances: The Basic Tool for Energy Planning', in H. Neu and D. Bain, eds, *National Energy Planning and Management in Developing Countries* (D. Reidel Publishing Company, 1983), 203.

governance and planning to developing countries in the 1980s. By the 1990s, energy flows and conversions worldwide had been brought into a common form, which consultants, policymakers, and investors had learned to read.

The WEC was one of the first to pick up the idea of a world balance in 1975. It perceived the oil crisis as a mission 'to produce a world plan for energy', which meant the compilation of a global energy balance including developing countries to forecast energy demand.[97] There was a division of statistical labour between the OECD and the WEC: The OECD (and its suborganization IEA) would focus on the statistics required for short- and mid-term decision-making, while the WEC would compile more long-term world balances.[98] The Conservation Commission, formed to that end, set up its headquarters in the IAEA in Vienna, not far from the International Institute for Applied Systems Analysis (IIASA), from where it published a series of energy scenarios over the following decade. In its early years, the global mission was not reflected in the composition of the commission, whose members came almost entirely from the US and Western Europe. Acting from an understanding that the world economy was a common good to which every country was expected to contribute, the WEC deemed the inclusion of developing countries necessary as they 'often dispose of a large amount of the resources required to meet the world's needs'.[99] The

97 World Energy Conference, 'Minutes of the Meeting of the International Executive Council, Held in the Falconer Centret, Copenhagen, Denmark, on Monday, 26th May and Wednesday, 28th May', 1975, Annex 4. A second 'world energy balance' was calculated around the same time by the Workshop on Alternative Energy Strategies at the Massachusetts Institute of Technology; C. L. Wilson, *Energy: Global Prospects, 1985–2000 [Report of Workshop on Alternative Energy Strategies]* (McGraw-Hill, 1976). Both coupled energy balances with econometric-demand models.

98 World Energy Conference, 'Minutes of the Meeting of the International Executive Council, Held in Cobo Hall, Detroit, Michigan, USA, on Friday, 20th September, Saturday, 21st September and Friday, 27th September', 1974, Annex 4.

99 This echoes the mandate era's understanding, 'which characterized the resources of the mandate territories as somehow belonging to humanity as a whole'; Antony Anghie, *Imperialism, Sovereignty and the Making of International Law* (Cambridge University Press, 2007), 212. World Energy Conference, 'Minutes of the

first report published in 1978 was cautiously optimistic that the world could shift its energy demand from oil to nuclear power and coal in sectors where this was possible. To do so, governments should bring about the conditions under which the energy-supply industry could thrive and assure 'that the long-term utilisation of all forms of non-renewable resource will be supported and that investment risks will be minimised'.[100]

In globalizing energy balances, it became apparent how much the method had always relied on commodified energy. Only in the form of a commodity, as a thing that was bought and sold, imported and exported, did nature's work leave traces on accounting sheets. However, a large part of world population did not buy fuel in a formal market exchange but relied on locally collected wood, vegetable waste, and dung. While the WEC's Conservation Commission prided itself on having forecast global energy demand in some detail for the first time, its modelling team found no better way to deal with non-commercial energy than to hold it constant even though it amounted to an estimated 13 per cent of world total.[101] The model could not account for the logics that governed over a tenth of global energy demand, as quantitative methods were helpless when dealing with traditional fuels.[102] Future reports by the Conservation Commission adopted a more qualitative approach. Regional balances were constructed from the ground up, and experts from Global South countries estimated non-commercial fuel use.[103]

International Executive Council, Held at the Palais des Congrès Houphouet-Boigny Abidjan, Ivory Coast, on Monday, 11th October, Tuesday, 12th October, and Wednesday, 13th October', 1976, 48. In its later work under former EDF manager Jean-Romain Frisch, the commission's work became more collaborative, hoping to inspire a self-reflection within developing countries about their own energy futures.

100 World Energy Conference, *World Energy: Looking Ahead to 2020* (IPC Science and Technology Press, 1978), 11.

101 Ibid., 198, 201.

102 R. K. Pachauri, 'Third World Energy Policies: The Urban-Rural Divide', *Energy Policy* 11, no. 3 (1 September 1983): 220.

103 Jean-Romain Frisch and World Energy Conference, *Energy 2000–2020: World Prospects and Regional Stresses* (Graham & Trotman, 1983).

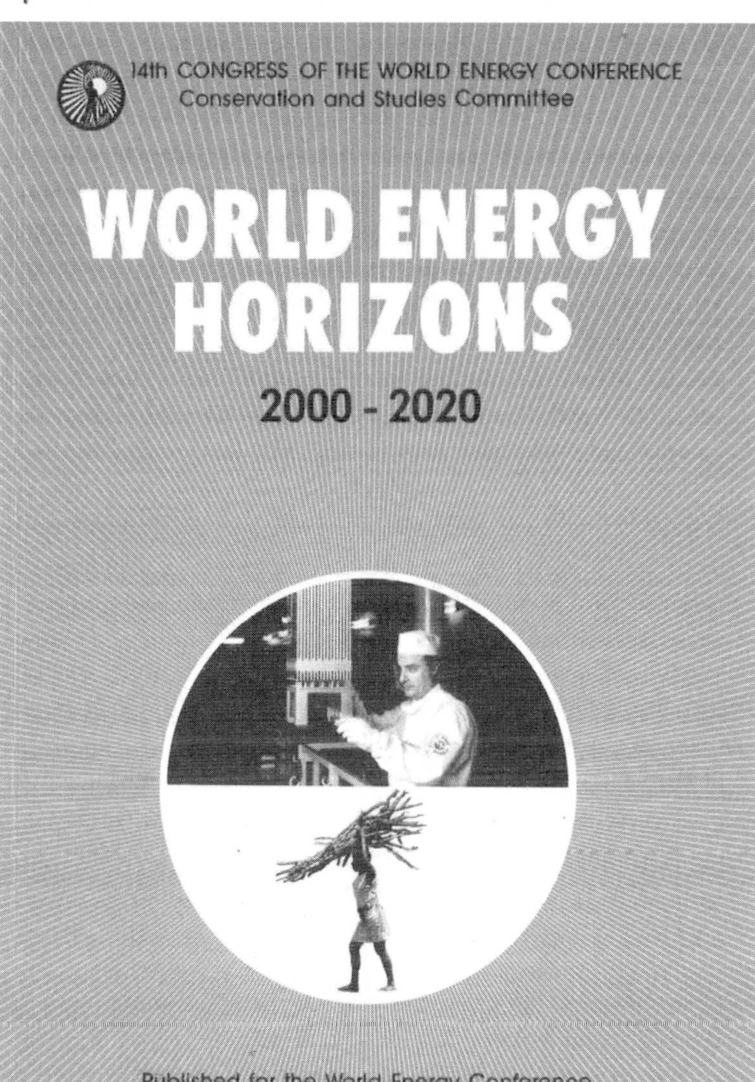

Figure 5.3. *World Energy Horizons*, the fourth report of the WEC's Conservation Commission, published in 1989

These numbers on non-commercial energy evoked different futures to different observers: A better estimation of fuel and dung use, and its possibilities of substitution, could serve a developmental purpose. With growth slowing in the OECD countries, it could be seen as an untapped market. As a demand that would, in future, potentially be satisfied by petroleum products, however, it could also be understood as a danger to the world energy situation. Whether welcomed or dreaded, however, non-commercial energy use was explicitly understood as *commercializable* demand.

Energy balances are the model in which the energy economy and its futures are today imagined. Like all models, they risk being mistaken for reality and become untrue to the object they 'represent' in a formalized way. A condition of their possibility is the historical development of a space for substitution, particularly through electrification and a diversification of fuels. Yet the commensuration of energies in balances exceeds the substitutions that take place at any time; they create a statistical equivalence that goes beyond what is technically possible and a comparability of uses that neglects their particularity. There is a politics of methodology involved in energy balancing. Decisions on what should count as an energetic process, how energies should be valued in present and future, and how energy uses would be differentiated, affected how the energy economy looked like. Such questions determined the way in which industries were represented in the balance: their size in relation to each other, their contribution to the nation's energy supply, and the future open to them. The chosen methodology (the limits, sectors, conversion factors, the categories of final energy consumption, etc.) decides which differences become visible in the model. They are not the only differences that matter – but they are the differences that matter for the political and economic practice those balances inform.

An energy balance is an instrument of governance and regulation. It serves to manage, plan, and study energy commodities in their interrelation and mutual dependence, covering their transformations both along the value chain from 'raw energy' (e.g., coal) to 'produced energy' (e.g., electricity) and 'useful energy' (e.g., heat) and across markets – that is, covering all energy products that fulfil a particular function (e.g., generate electricity, supply heat). Such balances might first seem like an engineer's point of view on the national economic space. However, I have argued that this abstraction was rooted in the problem of investing in the coal, oil, gas, and electricity sectors at a time when their future – and, thus, future revenue – had begun to depend on each other. Balances reflect the view of capital and the capitalist state to compare the costs of units of energy across sectors and the development of sectors in relation to each other. While it has its roots in the problem of marshalling cheap energy for economic growth in the postwar years, it only became global with the problem of oil substitution in the 1970s and the ensuing liberation of energy markets and global investment in the 1990s. The worldwide standardization of energy balancing means that the size of a country's heating market, the potential for transport electrification, or the state of decarbonization can be gleaned from a balance sheet.

Energy balancing has informed a wide range of policies – from investment planning to conservation efforts, and climate mitigation. As we have seen in the discussion among engineers, it once offered a space of experimentation in which different 'energy economies' could be modelled and compared. While methodologies today still differ, this mostly refers to the level of recalculation (such as conversion factors), not to the creation and inclusion of different energy data or to a different typology of uses, both of which would require much more data-gathering effort.[104]

104 The IEA adopted the substitution method for its early balancing work. However, it had to abandon it for primary energy input after 1990, when energy balances were harmonized globally. The method benefits the electricity industry as it values electricity higher than fuels. It also carries on a certain historical perspective on the energy problem, one born out of the political project of oil substitution, which sees everything in relation to fossil fuels. The IEA now argues that the substitution method was relevant only for countries with a small share of water power. With long-term energy balances, as they are needed for climate modelling, also comes the practical problem of weighting renewables and nuclear power according to the changing efficiency of thermal power plants. Conversion factors would have to be historically specific and would vary considerably between 1890 and today. The IEA uses now the 'physical energy content' method, which assigns to water and solar power a 100 per cent efficiency. This overblown efficiency, however, reduces their primary energy input and can make it appear small in relation to that of fossil fuels. Erik Sauar, 'IEA Underreports Contribution of Solar and Wind to World Energy Mix', *Energy Post* (blog), 31 August 2017.

Government agencies no longer experiment with the method, as they did during the 1950s, when engineers developed a wide range of balances, including an exergy balance.

Balances mediate identity and difference of global energy products. Against the background of an assumed identity and equilibrium, the method brings out a limited number of differences that are relevant to govern the reproduction of the system. This begs the question whether we could repurpose this method that came up during the politics of economic growth: By changing what is deemed a relevant difference, could we govern the energy economy during a time of decarbonization? Balances could be used to track the ownership of the energetic forces of production and to model private or public oversight over energy conversions – if only we had the power to collect and assemble this data. In a variation of the Soviet balance distinguishing between sectors of production and reproduction, one could make a difference between fossil fuels used for building a non-fossil-fuelled infrastructure and fossil fuels burned for mere reproduction of the old. However, this would not free us from the need to search for the particularities that are overlooked when all energies are treated as economically equal.

Conclusion

Energy and Emancipation

What is unchangeable in nature may take care of itself. Our task is to change it.

– Theodor W. Adorno, 'Reaktion und Fortschritt'

At the first World Power Conference in 1924, a British colonial engineer reported on the Gold Coast's potential power resources, recalling the moment when the landscape presented him with an opportunity: 'During a rapid canoe voyage down the river in 1915, I noticed on passing through the narrow gorge near Ajena that it was an ideal place for a dam.' Alas, the river's strong current allowed only for a quick glimpse, and 'there was no opportunity to take measurements, nor has there been any since'.[1] Plans were drawn up only in the 1940s. Two decades later, the first prime minister of independent Ghana, Kwame Nkrumah, adopted the project hoping the hydroelectric plant would power pan-African development and economic independence. Today, standardization of units and markets, global energy balances, and maps of energetic

1 World Power Conference, ed., *The Transactions of the First World Power Conference, London, June 30th to July 12th, 1924* (P. Lund, Humphries & Co., 1925).

potential make it easier to seize energetic opportunities from afar. To 'harness nature' – to control it for use of power – requires a vision, experience, and knowledge. This book can be seen as a history of the energetic point of view, a perspective that becomes possible only in a certain kind of society. The energetic viewpoint is modern in the sense that nature appears for it no longer as a force of resistance nor as power exceeding the capacity of human technology, but as an invitation to technical improvement itself. A gorge raises the prospect of a dam in the mind of the electrical engineer. Nature appears to be calling for human reins and bridles, as Hans Blumenberg put it.[2]

Historical materialism contains a trace of this modern concept of nature. Historians and social scientists working on capitalism's ecological contradictions are likely to pause when they encounter Adorno's line that 'what is unchangeable in nature may take care of itself. Our task is to change it.'[3] This call to change nature sounds unbearably modern. Does it not express the very ideology that nature is not enough, that it is separate from us, and that we must change and improve it? Is it not an affirmation of the energetic point of view chronicled here – to harness a maximum of nature's work to change nature? Furthermore, does it not deny any environmentalist critique referring to fixed characteristics of nature? Insurmountable natural limits, if they exist, can take care of themselves. And, lastly, does this not imply that we are not nature, that our task is to speak on behalf not of nature, but of ourselves in and against nature?

In Adorno's thought, nature is the obverse not of society but of history. History is that in which the new appears; nature is the experience of that which reproduces itself blindly, of that in which the subject cannot recognize traces of its own; nature is the alien, the cryptic, the

2 Hans Blumenberg, 'Phenomenological Aspects of Life-World and Technization (1963)', in Hannes Bajohr, Florian Fuchs, and Joe Paul Kroll, eds, *History, Metaphors, Fables: A Hans Blumenberg Reader* (Cornell University Press, 2020), 359.

3 Theodor W. Adorno, 'Reaktion und Fortschritt', in *Musikalische Schriften*, vol. 4, *Moments musicaux impromptus* (Suhrkamp, 2015), 139.

mythical.[4] Under capitalist alienation, modern society has become natural in this sense; it has become a second nature, because 'the law of capitalist accumulation has been mystified into a law of nature'.[5] There is a lack of history and an abundance of nature in the Capitalocene. When Adorno speaks of 'nature' in the quote above, he refers to that which has come to reproduce itself in a quasi-natural way under capitalist conditions. Note that the 'unchangeable' is defined only negatively here: Adorno does not say *what it is that is unchangeable* in nature. This is because it cannot be given positively as a list of natural laws and social structures by the very subject that is at the same time *subjected* and conditioned by them. What is unchangeable is neglectable to those looking for change: It remains in the dark background of what conditions human life or will make itself felt as resistance and as such remain a moment of human activity. First and second natures have a way of caring for themselves. What is affirmed here, then, is not the mastery of external nature but historical consciousness – an attitude towards the past that seeks to emancipate itself from the past by groping for the changeable, for a thought that can serve an anticipated culture.

There is a second aspect in which Adorno's focus on change speaks to the ecological question. Human societies have long viewed nature as a source of social norms and orders. In doing so, they have often projected their own values onto nature. If human history is understood to take place in an energetically ordered nature, this implies an economics and prescribes a logic of individual and social development. François Vatin describes energetics as an economy of machinery that implies decision pertaining to value. In conceiving of machines as a system for the transformation of energy, engineers established a norm of value (work) and a related measure of efficiency to determine the meaning of progress. Yet even the strict, narrow problems of engineering escape such a one-dimensional reduction. The search for an unchangeable order in nature

4 Theodor W. Adorno, 'The Idea of Natural History', in Robert Hullot-Kentor, ed., *Things Beyond Resemblance: Collected Essays on Theodor W. Adorno* (Columbia University Press, 2006), 252–69.

5 Theodor W. Adorno, *Negative Dialectics* (Routledge, 1973), 354.

onto which society can be based and to which it must conform can be repressive insofar as it reduces individuals to material beings.[6] It also quashes the experimentation with other social orders that can be realized in nature but are yet unknown.

In this book, the energy economy, the spheres of production, circulation, and consumption of nature's work, has been treated as a second nature. This has meant, first, recognition of the energy economy as *historical* fact: It developed neither as a result of natural human needs nor out of natural energetic laws. Rather, the objectification of natural forces set in train by capitalist industrialization brought something new into the world through the manipulation of natural things and human bodies. By increasing productivity, industrialization also opened the possibility of human life without suffering. But the employment of nature's work to improve the conditions of human life has remained restricted, fragmented, and ambiguous. Energetic forces of production are instrumental in maintaining class society. Second, the appropriation of nature's work continues objectively, and therefore it takes on a *natural* character. Recall the Planet-Generator, sitting in the void: How much of nature's work was and is still spent to preserve the status quo? The twenty-two zettajoules expended since 1950 stand in no relationship to real emancipation: This power and the fact that human suffering persists are increasingly contradictory. With climate change and militarization, the destructive aspect of energetic forces will become only more pronounced. Changing society's relation to nature more broadly remains a worthy project.

Earlier projects dealing with energetic emancipation tended to identify progress with the abstract potential of 'change' produced or sold within the energy economy. We have encountered attempts to appropriate the energy economy for emancipatory purposes along such lines. Bogdanov embraced energetic abstraction as the 'labour point of view'

6 Lorraine Daston, *Against Nature* (MIT Press, 2019); François Vatin, *Le travail, économie et physique*, 1st ed. (PUF, 1993), 125–7; Theodor W. Adorno, *Philosophische Terminologie: Zur Einleitung*, vol. 2 (Suhrkamp, 1974), 187–8.

from which the world could be organized according to the needs of the proletariat, kilowatt-hour for kilowatt-hour. To interwar socialists, electrification offered a way not only to 'socialize' the energetic means of production but to realize a new society mediated through oscillating electrons. Matters of organization, rationalization, and ownership were important, though they were often justified primarily with their effect on higher productivity without focusing on how this would improve people's lives. Yet one should not forget that the Soviet Union, and later postcolonial states, assumed (not without reason) that their survival depended on accelerated industrialization. In pitting the energetic perspective against capitalist and imperialist states, such projects tended to identify a socialist society with a 'republic of labour' organized along the lines of energy. But even if reducing human suffering and combatting imperialism can be achieved through developing energetic forces, it would be wrong to conclude that emancipation grows in units of energy.[7]

With climate change, visions of an improved energy economy often focus on changes in technology and a substitution of one form of energy for another. In short, they amount to an 'energy transition' from high-carbon fossil fuels to low-carbon solar and wind power. Capitalism subsists on fossil fuel; will it die once such aliments are depleted? Could a decentralized, flowing nature of renewables catalyse a different kind of social organization and a new relation to nature? Hermann Scheer's hopes for global solar energy democracy reiterate the energy-historical trope that new forms of energy might revolutionize society. Along similar lines, Elmar Altvater framed the natural contradiction of capitalism in thermodynamic terms and tended to identify the 'capitalist energy system' with wasteful fossil fuels. Today, with the formation of a green

7 Walter Benjamin, *Illuminations: Essays and Reflections* (Schocken Books, 1968), 258–9; Elizabeth Chatterjee, 'The Colony and the World Energy Revolution: Meghnad Saha's Energetic Developmentalism', in Daniela Russ and Thomas Turnbull, eds, *Energy's History: Toward a Global Canon* (Stanford University Press, 2025), 84–102; Daniela Russ, 'The Red Thread to Socialism: Gleb Krzhizhanovskii's "Energetics and Socialist Reconstruction"', in Russ and Turnbull, *Energy's History*, 103–23; Stephan F. Miescher, '"Nkrumah's Baby": The Akosombo Dam and the Dream of Development in Ghana, 1952–1966', *Water History* 6, no. 4 (December 2014): 358.

capitalist class and dense linkages between technology and energy sectors, it seems implausible to argue that change will arrive automatically with an expanded use of renewable energy sources. The energy transition is not taking place fast enough, and renewables are hardly democratizing the energy sector.[8]

Both productivism and the energy-transition fetish are dead ends. This book has presented the history of the energy economy as an ongoing development of the energetic forces of production that persists in the face of social and natural resistance. Originating in the objectification of the motor mechanism, this subsumption soon went beyond mechanical work and affected the various ways in which nature can be put to work. Thermal, electrical, chemical, and informational transformations are now all comparable, can be recalculated into each other, and are treated as forms of 'pure change in time'. This ability to produce a constant amount of controlled, flexible, and scalable change in production constitutes the use value of energy commodities. From steam engines to electrification and digital technologies, the control and manipulation of nature's work remained tightly linked to the disciplining and substitution of a source of work that was more difficult to control: human labour.

This capitalist socialization of nature intensifies and expands but is never total. It yields a combined and uneven development of the global energy economy, where commodified and non-commodified fuels coexist, and renewable and non-renewable sources are frequently combined to smooth out each other's disadvantages.[9] What is more, socialization cannot subsume nature: It cannot control the conditions of its own possibility.[10] By harnessing nature's work and selling it off in pieces, energy capitalists do not control the reproduction of that capacity to work, nor do they exhaust or grasp the ways in which nature changes, or

8 Hermann Scheer, *The Solar Economy: Renewable Energy for a Sustainable Global Future* (Earthscan, 2002); Elmar Altvater, 'The Social and Natural Environment of Fossil Capitalism', *Socialist Register* 43, no. 43 (2007): 37–49.

9 In this sense, it fits well with the 'symbiotic' history of energy offered by Jean-Baptiste Fressoz, *More and More and More: An All-Consuming History of Energy* (Allen Lane, 2024).

10 Deborah Cook, *Adorno on Nature* (Routledge, 2014), 23.

will change, in its entirety. Just like abstract labour preserves a part of the worker's subjectivity in negating it, the concept of energy, by negating nature's manifold transformations, conserves something of its objectivity. In this sense, capital's forms are always in and against nature.

From this perspective of an always incomplete subsumption of natural forces, the link between capitalism and fossil fuels becomes brittle. If only the congruence of fossil fuels and capitalism existed, if their only property were being 'abstractable', the structure and natural contradictions of the fossil industry could not be understood. Fossil fuels are concentrated, mobile, and energy-dense. But they have many other features that do not lend themselves to being exploited, as we have seen in the chapter on coal. The most glaring 'negative' property of fossil fuels is the contradiction that affects all extractive industries: They do not naturally reproduce at the speed of capital's infinite self-valuation and capital cannot technically control the conditions of their reproduction. Another is the widely varying, and declining, quality of stocks, which affects production costs and conditions how companies and oil-producing states hedge their bets in petroleum's endgame.

A more deeply electrified economy has developed within, not against, capitalism. While some electric capitalists hold fossil assets (historically, this could mean a coal mine; today, a thermal power plant), at the point of consumption, they have competed with fossil capital, and electricity has gradually replaced the direct use of fossil fuels. Today's wave of electrification of heating and transport, along with the electrification of tasks that comes with digitalization, could be interpreted as one moment in a process that could lead to an (almost) all-electric capitalist energy system. Electricity is not only the most flexible energy commodity but also the raw material of the data economy. Apart from the old electric capitalists, such as Siemens or General Electric, electrification is backed today by a new and powerful capital faction: Google, Amazon, Microsoft, and Meta have already invested in electric assets – particularly wind and solar power, and now potentially nuclear power – at a higher rate than in other energy sources. At the same time, tech companies are invested in a range of grid technologies, such as smart meters, AI to optimize energy usage, or the coordination of distributed producers through virtual power plants. In the US, Google already acts as a utility, while Google, Microsoft, and Amazon urge Ireland to allow them to construct private power lines connecting data centres with power plants. Whether it is fossil or techno-electric capital that defines the futures of the energy economy remains to be seen.

These new electric capitalists would not phase out fossil fuels to fight climate change either, but they also have no particular interest in them. As owners of the means of electricity production and distribution, they have an interest in the electric system and the growth of electricity consumption, not in single resources. Still, they have benefited from decarbonization policies which have expanded electrification, allowed them to enter a formerly oligopolistic market through subsidies, and created a higher need for grid services. The 'electro-state' willing to push for a mineral-based energy system might quickly de-risk the financing of renewable energy infrastructure and new mineral exploitation and guarantee profits through subsidies.[11] In this context, the physical and digital engineering of an intermittent flow of energy into a stable source of power is already underway. Market designers are busy inventing regulations to counter seasonal and weather fluctuations so that solar and wind power will remain profitable. Chinese engineers have developed ultra-high-voltage (UHV) transmission lines enabling wind and solar power to be sourced from continental and ocean-sized zones, to be distributed over a continents-spanning grid. Yet even this global grid is expected to require roughly 30 per cent dispatchable power sources, meaning fossil, nuclear, or hydrogen fuel, to keep power flowing.[12] Electricity could be stored in the digitally connected batteries of prosumers, which would be fully coordinated by an AI owned by a

---

11 Brett Christophers, *Our Lives in Their Portfolios: Why Asset Managers Own the World* (Verso, 2023), 125; Brett Christophers, *The Price Is Wrong: Why Capitalism Won't Save the Planet* (Verso, 2024); Daniela Gabor and Ndongo Samba Sylla, 'Derisking Developmentalism: A Tale of Green Hydrogen', *Development and Change* 54, no. 5 (2023): 1169–96; Jana Tauschinski, Nassos Stylianou, and Edward White, 'How Xi Sparked China's Electricity Revolution', *Financial Times*, 12 May 2025.

12 Cong Wu, Xiao-Ping Zhang, and Michael J. H. Sterling, 'Global Electricity Interconnection with 100% Renewable Energy Generation', *IEEE Access* 9 (2021): 113169–86; Zhenya Liu, *Global Energy Interconnection* (Academic Press, 2015).

tech company that is both the largest producer and consumer on the grid. Then, finally, capital's power source would be as infinite as its accumulation drive, only limited in space by the surface of the earth and in time by the sun's slow burn-up.

Yet it would still be impossible to compel nature to conform entirely to such visions. When fuel scientists had to deal with a complex material formed in natural history, they were dealing with the past. Scientists calculating wind and solar power must concern themselves with a natural phenomenon that may change yet, due to natural and social causes. A renewable energy system would be more reliable on weather-dependent sources at a time when weather is expected to become less predictable. It would have to deal with more frequent extremes and the uncertainty that weather patterns could change with the climate, as can be seen in the debate on 'global terrestrial stilling'.[13]

Perhaps the strongest argument against the likelihood of an electrified world running largely on renewables is geopolitical: Fossil fuels and electrical grids undergo abstraction from concrete space quite differently. The latter only transcend space by means of infrastructure requiring constant, instantaneous social coordination. Whether this coordination in a system-operating agency is considered sufficiently isolated from politics is dependent on the severity of political tension. As a rule, the higher the share of wind and solar power, the larger the area from which power would have to be sourced. By contrast, barrelled petroleum will find its way through even the most fragmented global space.[14] As long as the ability to wage war depends on fossil fuels, will states ever demolish fossil infrastructure? And how are continent-spanning grids to work amid a fragmentation into regional political

13 Gan Zhang, 'Amplified Summer Wind Stilling and Land Warming Compound Energy Risks in Northern Midlatitudes', *Environmental Research Letters* 20, no. 3 (February 2025): 034015; David E. H. J. Gernaat et al., 'Climate Change Impacts on Renewable Energy Supply', *Nature Climate Change* 11, no. 2 (February 2021): 119–25; Dongsheng Zheng et al., 'Climate Change Impacts on the Extreme Power Shortage Events of Wind-Solar Supply Systems Worldwide During 1980–2022', *Nature Communications* 15, no. 1 (18 June 2024): 5225.

14 Charles Perragin and Guillaume Renouard, 'The Shadow Fleet That Keeps Russian Oil Flowing', *Le Monde diplomatique*, 1 March 2025.

blocs? Such possibilities remind us that any emerging electrical capitalism will face daunting obstacles.

What is at stake in recalling nature's resistance, if it has not stopped, and will not stop, capital's global employment of natural forces on human time scales? To highlight this fact is not equivalent to a claim that capitalism will fail on natural grounds – there is no need to read 'contradiction' in a teleological sense as something that anticipates its own overcoming. Rather, emphasis on non-identity is crucial in salvaging the 'materialism' of historical materialism and to caution it against idealisms masquerading as materialist, energetic, or otherwise. Part of this endeavour requires reading the idle, unemployable, resistant nature – the residuum of nature's work – as evidence that radical change in our relation to nature is possible.

Some in the humanities see the concept of energy as a modern abstraction and thereby tainted. Energy from this perspective is a reduction of the various qualities of natural things to one common denominator. Philip Mirowski is particularly withering in his discussion of the energy abstraction. It is 'a specific world view, one bound up with Laplacean determinism, temporal symmetry, a reductionist conception of the body, and the exchange of equivalents'.[15] On Barak, for his part, calls it a 'fossil-fuel based abstraction', which 'directed attention away from the features that made [different powers] specific, indeed, from their very materiality'.[16] For Cara Daggett, energy is a 'geo-theology . . . that combines the prestige of physics with the appeal of Protestantism' and reinforces 'our commitment to growth . . . in order to support the interests of an industrial, imperial West'.[17] Through the concept of energy, society grapples with nature's ability to perform work and disregards all other natural or social

15 Philip Mirowski, *More Heat Than Light: Economics as Social Physics, Physics as Nature's Economics* (Cambridge University Press, 1989), 133.

16 On Barak, 'Three Watersheds in the History of Energy', *Comparative Studies of South Asia, Africa and the Middle East* 34, no. 3 (1 January 2014): 440–1.

17 Cara New Daggett, *The Birth of Energy: Fossil Fuels, Thermodynamics, and the Politics of Work* (Duke University Press, 2019), 190.

relationships in which human beings and natural things stand. In doing so, it expresses a diminished and impoverished experience of nature and historical change.

The violence of this abstraction is particularly notable when applied to the human body or pristine elements of nature. From the energetic worldview, the human body is a machine – and the machine is a 'measure of men'.[18] The calculation of 'energy' can then be instrumentalized to reduce wages to a subsistence level, to keep prisoners barely alive, or to starve populations.[19] One may also experience its effects in new landscapes drawn into the valorization of capital, as they are refashioned in the name of energy production. Where the proliferation of electric cars is tied to net-zero goals or the rising capacity of lithium mines, the concept of energy mediates the terms. From the energetic vantage, mountains, rivers, and forests appear principally as sources of power and are severed from their local contexts. Given the concept's reductive abstraction, its implication in the exploitation of nature and labour and its teleology, some argue that historians and social scientists must be wary of using the concept of energy at all.[20] Does progress speak no longer the language of energy?

If the energy concept is rooted in the development of productive forces, as has been argued here, it cannot so easily be treated in isolation from the reproduction of human life. The concept is today both reified and objective. Reification has objectified it in a double sense, as related to both first and second natures. First, the energy concept expresses the truth that society lives not from its own powers. The control of nature's work has in fact increased society's dependence upon nature. In negating the manifold ways in which nature changes, energy still preserves

18 Anson Rabinbach, *The Human Motor: Energy, Fatigue, and the Origins of Modernity* (Basic Books, 1990); Michael Adas, *Machines as the Measure of Men: Science, Technology, and Ideologies of Western Dominance* (Cornell University Press, 2015).

19 Nina Mackert, 'Feeding Productive Bodies: Calories, Nutritional Values, and Ability in the Progressive-Era US', in Peter-Paul Bänziger and Mischa Suter, eds, *Histories of Productivity* (Routledge, 2016), 117–35.

20 Barak, 'Three Watersheds in the History of Energy', 151; John Urry, 'The Problem of Energy', *Theory, Culture and Society* 31, no. 5 (1 September 2014): 5.

something of its changeability. Through the energy concept, society reflects on first nature. Second, this is also true in the sense that all social reproduction is today mediated by energy. This is the meaning of second nature: The capitalist energy economy confronts individuals as an underlying, pregiven, resistant structure reproducing itself behind their backs. Within this second nature, it is difficult to imagine how an entirely different relationship to nature could come about. It certainly does not, as we have seen above, simply coincide with a transition to new 'sources of energy'. If the use of renewables can be said to conform more closely to the reproduction of the planet's energetic potential, this adaptiveness is mediated through the most reified form of energy, electricity. More dubious still is the idea that our energetic predicament might change by *thinking* the society–nature relation differently: Unaffected by a change in social theory, the energy concept will continue to mediate social reproduction.

In a remarkable part in *Negative Dialectics*, Adorno defends reification against what he sees as a philosophers' trick of overcoming it by means of new categories. 'In relativizing or liquefying that thingness', argues Adorno, 'philosophy believes to be above the supremacy of the commodity form, and above the form of subjective reflection on that supremacy, the reified consciousness.'[21] Focusing criticism on reification is a misguided opposition to the existing world, its objectivity and historical meaning. Nature and society have diverged and should be reconciled, but this reconciliation cannot take the form of immediate unity, and neither does it represent balance and harmony with nature. Because divergence is not a matter of theory but is objectively sustained in social reproduction and sedimented in subjectivities and institutions, no immediate reconciliation can take place.[22] Can the reification entailed by the concept of energy nevertheless be defended as a component of a more emancipatory relationship to nature?

21 Adorno, *Negative Dialectics*, 189.

22 Adrian Wilding, 'Naturphilosophie Redivivus: On Bruno Latour's "Political Ecology"', *Cosmos and History: The Journal of Natural and Social Philosophy* 6, no. 1 (11 September 2010): 31.

Critique of reification and abstraction neglects the progressive aspect of the concept of energy. Along with the natural sciences, the reification of nature – not only, but also in the form of energy production – brought about the potential of a world without want.[23] The concept of energy made it possible to imagine and propagate the material and social development of societies, and a future society where labour would be organized more democratically. All works on the energy economy must grapple with how the destructive and productive often appear together, grow out of each other, as do the conditions of abundance and repression. Today, there are still traces of the progressive in the destructive. Fossil fuels undermine and enable their own transcendence simultaneously; they condition the use of nature's work for the purpose of human emancipation. Materially, a renewable economy will never be fossil-free; it will be a post-fossil economy, into whose technologies and infrastructures the use of fossil fuels has congealed. Renewable energy technologies themselves emerged from an energy economy that could command vast quantities of heat in order to generate silicon crystals, bake cement, and forge steel. Fossil fuel use is recorded in the carbon dioxide levels of the atmosphere, just as it is evident in physical energy infrastructure.

The energy concept allows a reflection on the objective state of society. We need categories for critical analysis of the reified relationships in which we live.[24] Where relations among people, and between society and nature, are mediated by energetic objects, the concept of energy becomes critically necessary. Such immanent critique, one levelling the concept of energy against the social system that engendered it, has a long history. Amy Wendling has argued that Marx did exactly this: By putting forth an energetic calculation of labour power opposed to capitalist calculations of value, he was able to identify surplus creation and exploitation. Likewise, the concept of energy is used in theories of ecologically unfair exchange to reveal that apparently symmetric market exchange 'disguises an asymmetric net flow of embodied matter, energy

23 Adorno, *Negative Dialectics*, 189.

24 Reification is the form in which false objectivity can be reflected. See ibid.

and labour from the periphery to the core'.[25] The energy concept itself thus indexes the non-identity between what capitalism pretends to be, and what it really is.

Adorno reads in the impulse to 'liquefy' reification a certain hostility to otherness – it is the subject's effort at overcoming the object. Rather than denying energy's progressive aspects then, one might ask what constitutes a good relation to the object, understood as a prerequisite and mediation of a good social praxis. Might there be a reconciliation – not one that grounds itself in a natural order, but one that strives for another relation between people and energetic things – achieved through historical awareness? The most reified form of energy, the electrical current, has historically sparked the boldest visions of emancipation through the social organization of energy. The grid allows for the imagination of a complementary relationship, the realization of energetic independence in social dependence. What would electrical engineering look like if it were guided by a sense of the good life and the socialization of the energetic means of production – not against but around the peculiar materiality of the electric current?

The energy concept would then be an instrument of critique, no yardstick of future societies. It could no longer be treated as the foundation on which nature and society can be reconciled. A foundational perspective on energy is still very much alive, when global energy models are taken as proof that reality can and should be adapted to the solutions of the model or when claims to democratize the social surplus are countered with the natural limitedness of resources. From capitalism's energetic and material inefficiency, even from the immense destruction it wages through the energetic forces of production, does not follow that an energetically rational society would be desirable. The energetic

25 Alf Hornborg, 'Towards an Ecological Theory of Unequal Exchange: Articulating World System Theory and Ecological Economics', *Ecological Economics* 25, no. 1 (1 April 1998): 127–36; Oliver Schlaudt, 'The Market as a "Rigged Game": Theories of Ecologically Unequal Exchange and Their Implications for Value, Price, and Measures of Real Wealth', in Isabel Feichtner and Geoff Gordon, eds, *Constitutions of Value* (Routledge, 2023), 280.

perspective is neither natural nor neutral, and any attempt to organize society according to its logic – whether in the name of productivism or ecology – would risk suppressing incommensurate ways of life. By conflating Marx's definition of labour – the actions regulating the metabolism of humanity and nature – with a narrow calculation of energy, labour is thereby restricted conceptually to the specific form it takes in capitalism. Historical materialism should look for alternative conceptions of human beings as bearers of forces that are not their own.

Acknowledgements

The thoughts and words in this book have accompanied me as I have moved across institutional, disciplinary, and national boundaries over the past ten years. Many people have encouraged and supported me in various ways during this time – this is the place to thank them. The research and writing of this book were funded by the Mercator Foundation, the Fulbright Programme, the German Historical Institute in Moscow, the Fondation d'Électricité de France, the German Research Council, and the Remarque Institute.

The project began when I realized that sociologists studying the 'socio-ecological transition' were missing a concept of energy. During my time at the Forum Internationale Wissenschaft in Bonn, my colleagues' patience and persistence helped me transform this vague idea into a more tangible project. I am especially grateful to Julian Hamann, Evelyn Moser, and Katharina Kreuder-Sonnen. The institute would not have been the same without Raja Bernard's cheerfulness and solidarity. I could not have imagined a better place to complete my dissertation than the University of Bielefeld's Research Training Group 'World Politics'. Numerous conversations with Marc Jacobsen, Vera Linke, Johannes Nagel, and James Stafford helped me navigate the space between historical and sociological research.

I have been fortunate to have had stable funding, unwavering support, and unquestioned freedom while writing this book. For the first few years, I owe these exceptional research conditions to my supervisors, Tobias Werron and David Kaldewey, who supported me throughout the project despite all the detours I took. I am especially grateful to them for demonstrating that historical analysis still has a place in contemporary sociology. I would also like to thank Timothy Mitchell for supervising me during my time as a Fulbright fellow at Columbia University. As well as providing many helpful hints and suggestions, he impressed me greatly with the curiosity, kindness, and generosity with which he received my rudimentary dissertation project. Zaheer Baber was so kind to welcome me to the University of Toronto, where I began to seriously turn my dissertation into a book. My colleagues at the Institute for Global and European Studies in Leipzig – Steffi Marung, Megan Maruschke, Katarina Ristić, Karen Silva Torres, and Stephan Kaschner – enabled me to concentrate on completing the book when I returned to Germany with a half-finished manuscript. Thanks to Stefanos Geroulanos, a fellowship at the Remarque Institute at New York University in spring 2024 gave me the time and peace to finally complete the book.

Over the years, many people have helped me think through the arguments in this book, offered advice, and supported me in other ways. I have benefited countless times from Thomas Turnbull's erudition and generosity on all things energy-related, as well as his willingness to share and his curiosity to debate. Jacob Blumenfeld (seemingly) never tired of my questions on language and concepts, and introduced me to Chris Arthur's work. Without Andreas Malm, it would probably have taken me another year to finish this book. In fewer than a hundred words, Sebastian Budgen steered this book towards publication. Jacob Blumenfeld, Marco Vianna Franco, Rüdiger Graf, Matt Huber, Julia Kaiser, Moritz Klenk, Jan Overwijk, Hans Rackwitz, Florian Schmidt, Julia Schubert, Daniel Siegel, James Stafford, Thomas Turnbull, and Troy Vettese all read parts of the book and provided comments. Joshua Rahtz proofread and commented on earlier versions of most of the chapters.

Friendship has helped me through the despair that accompanies all thinking and writing in a bad world. I am grateful to Jacob Blumenfeld,

Simon Hecke, Julia Schubert, Thomas Turnbull, and Hannah Schilling for this. Above all, I would like to thank Moritz Klenk for his boundless enthusiasm and probing questions. Without him, this book would be a different one. Everything good in it was thought out together.

For shared excitement, for patience and distraction, and for our common pursuit of a single *Wissenschaft*, I thank Daniel Siegel.

Index